"In *Liberalism and the Challenge of Climate Change,* Chi tively, provocatively but accessibly, demolishes the cosy consensus that political and economic liberalism is capable of responding to the existential threat of climate change. With their emphasis on individualism, protecting the freedoms of capital, the primary of western scientific thought and faith in technological fixes, dominant liberal ideologies are having to confront their own crises and contradictions. This book expertly surveys and critiques these belief systems and imaginaries before exploring some of their contenders. It will be of interest to a range of students, scholars and practitioners working on climate change."

Professor Peter Newell, *University of Sussex and Research Director of the Rapid Transition Alliance*

"Christopher Shaw's essential and urgent book addresses the failure and fundamental inadequacy of current attempts to address the climate crisis. With disquieting clarity, he demonstrates how even well-intentioned participants in projects for preserving a livable planet are trapped within conceptual frameworks or paradigms that a priori prevent the emergence of meaningful strategies for averting catastrophe."

Jonathan Crary, *Meyer Schapiro Professor of Modern Art and Theory, Columbia University*

"Words fail us when confronted with the challenges posed by climate change. Deeds fail us as well. As Christopher Shaw demonstrates in this book, we are trapped in an ideological network spun by liberalism. This makes us blind to alternative and more radical ways of approaching climate change from a less individualistic and more communitarian perspective. This book should be read by anybody interested in understanding the climate change impasse in which the world finds itself. Understanding it is a precondition to moving beyond it."

Brigitte Nerlich, *Emeritus Professor of Science, Language and Society, University of Nottingham*

"Christopher Shaw is steeped in the sociology and politics of climate change. In this book he argues elegantly and powerfully across a range of areas that climate change is intertwined with liberalism and that this blocks any solution to the climate crisis."

Luke Martell, *author of* Alternative Societies: For a Pluralist Socialism

Liberalism and the Challenge of Climate Change

In this book Christopher Shaw analyses how liberalism has shaped our understanding of climate change and how liberalism is legitimated in the face of a crisis for which liberalism has no answers.

The language and symbolism we use to make sense of climate change arose in the post-World War II liberal institutions of the West. This language and symbolism, in neutralising the philosophical and ideological challenge climate change poses to the legitimacy of free market liberalism, has also closed off the possibility of imagining a different kind of future for humanity. The book is structured around a repurposing of the 'guardrail' concept, commonly used in climate science narratives to communicate the boundary between safe and dangerous climate change. Five discursive 'guardrails' are identified, which define a boundary between safe and dangerous ideas about how to respond to climate change. The theoretical treatment of these issues is complemented with data from interviews with opinion formers, decision-makers and campaigners, exploring what models of human nature and political possibilities guide their approach to the politics of climate change governance.

This book will be of great interest to students and scholars of climate change, liberal politics, environmental communication and environmental politics and philosophy, in general.

Christopher Shaw is Head of Research at Climate Outreach, and also holds the positions of Research Associate in the School of Global Studies, University of Sussex, and Director of DeSmog. Dr Shaw has worked in the field of climate change communication for over 15 years.

Liberalism and the Challenge of Climate Change

Christopher Shaw

Routledge
Taylor & Francis Group

LONDON AND NEW YORK

Designed cover image: Lachlan Gardiner / Climate Visuals

First published 2024
by Routledge
4 Park Square, Milton Park, Abingdon, Oxon OX14 4RN

and by Routledge
605 Third Avenue, New York, NY 10158

Routledge is an imprint of the Taylor & Francis Group, an informa business

British Library Cataloguing-in-Publication Data
A catalogue record for this book is available from the British Library

ISBN: 978-1-138-61504-5 (hbk)
ISBN: 978-1-138-61506-9 (pbk)
ISBN: 978-0-429-46348-8 (ebk)

DOI: 10.4324/9780429463488

Typeset in Times New Roman
by codeMantra

For Nicholas and Alex

Contents

Acknowledgements

I owe a huge debt of gratitude to colleagues past and present at Climate Outreach. The work does not represent the views of Climate Outreach.

There are a number of other people whose support, love and friendship have, in one way or another, helped me get this book writen. People like Peter Newell at the University of Sussex, who has given me innumerable hands up the precarious ladder of a career in climate change research when I have needed the support (Up the Albion!). To my compatriots in the Advancement Forum – our collective explorations have been an inspiration and a lifeline over the four years it has taken to write this book. Ben and Vicky, whose appetite for life and loyalty helps me remember what really matters. We lucked out there. Keith and Candice, who have never been anything other than kind, supportive and caring. And Ruth and Richard, the intellectual north star that guides my inferior efforts.

And finally, I wish to thank my family. My mum, Elizabeth, for the unconditional love. My brother Tim, and sister Louise, whose company is always a joy. Last, because not least, so far from least, Livi, my wife. You've believed in me all the way through, you have made all the right calls when it matters, and you continue to laugh at my jokes. Thank you.

Preface

This book makes a quite obvious point; the language we use and the stories we tell reflect our particular social and historical circumstances. The stories we hear about the solutions to climate change reflect the social and historical experiences of the liberal middle classes of the global North. They are stories intended to reproduce the privilege of the storytellers. I am writing from a subaltern position within the global North, that of someone who grew up in a family who struggled financially and a family with no experience of higher education. So, the middle-class world has always felt something of an 'other' to me. I am not of the middle-class world.

Not being of that world has provided me with an outsider's perspective on middle-class ownership of climate change campaigning and communication. This book is the viewpoint of someone stood at the window, looking in. I believe Cormac McCarthy once said that people write in lieu of blowing up the world. That feels an apt description of the motivation for this book. I am not a happy voyeur. Whilst I might characterise the middle classes as complacent in the normal run of affairs, and vicious when their privilege is threatened, I must also own up the anger that motivates the writing of this book. That anger is, I suppose, in some part the resentment of someone who has been turned away from the party, who feels not wanted. Yet also, the anger reflects the feeling of being lied to. Lies are easy to justify, easy to live with when you are the one doing the lying. Lies are more difficult to swallow as the one being deceived.

The lie I am talking about here is the lie that the middle classes of the global North are the carriers of the light, the owners of the world's future, the last word on what is good and right. My realisation of the extent and depth of this deceit became clear at the turn of the 21st century with the West's bombing campaigns against the poor of the world, across too many countries to mention. Today, the same class that promoted, justified and agitated for the killing of hundreds of thousands of innocent people is selling net zero as a just and effective response to climate change.

I have not been able to articulate an answer to our crisis. There are many ideas out there for how to 'save the planet'. Many of them seem to offer a quite attractive and reasonable vision of a climate-safe future. None of them, as far as I can make out, address how to take power out of the hands of those who are wrecking the world, and having wrested power from that class, keep it out of their hands. The other stories of our future do not recognise class and power as relevant issues and instead just demand everyone become middle class in a zero emissions, impact free, endless replay of existing social and political arrangements. My only contribution to the solutions literature is to note language is an important part of the problem we face, but only one part of it. Power is in large part material, physical. It seems as though the ability to ignore the political and create our own little individual paradise through consumption is simply too powerful to ever be replaced with the stomach for the fight needed to rescue ourselves.

I wish I could have paid more attention in this book to the role the global North/ South divide plays in blocking possible escape routes. The gendered nature of the climate crisis deserves more attention than it has received here. I have instead chosen to focus on language and class. Language, as part of an emancipatory agenda, can help build a better future. Language, as part of an oppressive and destructive agenda, can be the difference that confirms our desperate fate. We are stuck with the language of the latter situation. We must find a way to a new language, a language that can bring the subaltern classes of the world into the climate fight as part of an emancipatory and revolutionary agenda. This book contributes to that by helping explain where we are now in that journey.

Introduction

It is increasingly apparent that climate change will require a 'transformation of society' (United Nations Framework Convention on Climate Change [UNFCCC], 2017). The Intergovernmental Panel on Climate Change (IPCC), the body that synthesises climate science into policy relevant recommendations for governments, has concluded that

> targeting a climate resilient, sustainable world involves fundamental changes to how society functions, including changes to underlying values, world-views, ideologies, social structures, political and economic systems, and power relationships.
>
> (IPCC, 2022)

This book argues that the greatest barrier to bringing about these 'fundamental changes' is the stranglehold that liberalism has on our language, thoughts and imagination. What is presented as transformative climate action is actually action intended to legitimise liberalism in the face of a catastrophe for which liberalism has no answers.

There is little evidence that a majority of voters in the West have any desire for fundamental changes in social structures and worldviews. Even those who have brought into the need for fundamental and transformational change struggle to articulate what 'fundamental' change means, what it is that needs to change and how to bring about that change. It is as if liberalism, having delivered a profound increase in access to material goods, has exhausted itself, human imagination and the very possibility of living differently. Liberalism is now destroying the foundations of life by creating a world where limitless desires can be fulfilled through the marketplace, with access to capital the only constraint on consumption. The liberal prospect of a timeless future has given way to visions of no future at all (MacAskill, 2022).

All liberalism can offer in response to the encroaching biospheric terminus is technological tinkering to maintain 'the energy services essential to modern civilization' (Davis et al., 2018). Such a failure of political imagination, occurring within a philosophy intended to unleash the human imagination, demonstrates the intellectual bankruptcy of liberalism. In our post-political age, politics does not extend

DOI: 10.4324/9780429463488-1

beyond technical discussions of how best to administer liberalism. Extinction Rebellion and FridaysforFuture are keen to stress climate change has nothing to do with politics in the sense of values, worldviews and ideology. Instead campaigners should focus on ensuring policy-makers just listen to the science. Unfortunately, other than telling us climate change is happening and what impacts are expected when, the science has nothing to say on political responses to the crisis. After 250 years of reasoning and calculation ushered in by the rise of liberalism, the West has concluded that it is better to create a 'biological annihilation' that tears away 'the foundations of human civilisation' (Ceballos, Ehrlich and Dirzo, 2017), than even question the ultimate value of living in a society in which 'all social and institutional activity rests on free will and voluntary association' (Anderson, 1987: 37).

0.1   The promise of an emissions-free liberalism

Projections of a "safe" climate future are used to ground climate policy within the forms of knowledge and ideology that promote and reproduce the liberal worldview of elite actors from the global North (Knight, 2019). These climate projections also create a politically "safe" operating space for liberalism, by creating a timeline for action running to 2050. By granting permission to continue increasing atmospheric concentrations of greenhouse gases for (at the time of writing) the next 28 years, these agreements provide the political opportunity for 'business as usual' whilst scientists and engineers pursue the invention of new technologies that will make possible an emissions-free liberalism. The symbolism of climate 'guardrails' (WBGU, 1995, 1997, 2003, 2009, 2014) and 'planetary boundaries' (Steffen et al., 2015) are used to support claims of a knowable, universal and single level of acceptable climate risk. Statements such as 'scientists say we have to reduce greenhouse gas emissions to net zero by 2050 in order to avoid dangerous climate change' secures the terrain for narratives that promise a smooth and steady transition to a timeless and eternal emissions-free liberal utopia, a new heaven on earth.

The concepts of 'guardrails' and 'boundaries' are repurposed in this book. Instead of referring to safe levels of climate change, we identify and analyse the language and symbolism that are used as discursive 'guardrails' or 'boundaries' for safe ideas. These ideas, symbols, concepts and imaginaries are designed to keep the climate discourse away from promoting climate policies that are a threat to the hegemony of liberalism. The guardrails offer enough movement and space to create the illusion of free enquiry and open debate, whilst constraining policy development within the strictures of liberal market ideology.

The five climate guardrails around which this book is structured are:

- *Climate change is not a challenge to individualism.*
- *The liberal construction of climate change is universally true.*

- *Climate change is not an historical phenomenon.*
- *We have the technologies to solve climate change.*
- *New stories will save us.*

0.2 The liberal language of climate change

Climate change is 'that from which words fall back' (Shankara, cited Shah – Kazemi, 2006: 2). Unable to grasp the enormity of the crisis, we search for solutions in the vicinity of the problem (Anderson, 1987: 37). The vocabulary of 'liberalism' is the dominant and virtually all-pervasive idiom of our thought and speech in the West (Geuss, 2022). It therefore follows that the stories the West tells itself about climate change are actually stories about liberalism. Linguistically, this means using metaphors and analogies which draw on what is already familiar, to make sense of the unfamiliar in a process described as 'anchoring' (Moscovici, 1963). However, the speed and scale of anthropogenic climate, species extinction and destruction of the biosphere are so far outside of previous human experience (IPCC, 2021) that our stories of this phenomenon are unable to properly represent this reality (Bradshaw et al., 2021).

From the 1950s onwards it has been the scientists, policy-science institutions, media and environmental campaigners from the global North who have driven global awareness of climate change. These institutions are the products, creators and promoters of liberal capitalism and hence it is this ideology which has shaped the social reality in which scientific and social awareness of climate risks have arisen. The language used by these actors to make sense of climate change will necessarily be drawn from the lexicon of liberalism's symbolism and myths. Once these 'powerful and powerfully limiting frames have become embedded in the discourse they continue unchallenged' (Porter and Brown, 2001: 117). This results in a 'formidable system of exclusionary languages, gestures and judgements that serve to police the borders of debate' (Ross, 1991: 62). For those entering the climate policy arena today, it becomes difficult to see beyond these boundaries, to imagine that it is possible to talk of alternatives to free market liberalism (Fisher, 2009). Climate change research, policy and activism all share a common language of "transformation" and "fundamental change" to refer to a future which is politically, morally and economically exactly the same as the life-destroying liberalism of today.

0.3 Definitions of liberalism

'What is this liberalism that so shapes our lives?' ask Mann and Wainwright (2018: 80–81) 'Almost no one wants to define it – not even the liberals – and for good reason: it is slippery, contingent, blurry, dynamic and place and time specific' (ibid).

Nonetheless, we must stick our flag somewhere, and hereafter the term "liberalism" will be used to denote "bourgeois liberalism". Whether liberalism needs the qualifier "bourgeois" is questionable, given the 'metonymic identification of liberalism with the bourgeoisie' (Whyte, 1973: 324). The liberalism we are discussing is also synonymous with modernity (Spark, 2002: 12) and capitalism. This is a world where belief in God has been replaced with a belief in competitive private enterprise, technology, science and reason (Hobsbawm, 1988). Whilst socially and economically liberal (in order to ensure the whole population is equally exploitable and welcomed into the capitalist tent) bourgeois liberalism is politically liberal only to the extent that this political freedom does not challenge the hegemony of liberalism. The 'insulation of a politically instituted market from democratic politics qualifies this form of liberalism as authoritarian' (Streeck, 2016: 55).

The bourgeois element of the bourgeois liberalism being discussed in this book is, like liberalism itself, a contested concept with a chequered history. Moretti (2013) begins his analysis of this concept by noting that whereas "bourgeois" was once a very commonly used word, it is rarely mentioned today. Instead, the word has been replaced by "the middle class" (6–7). The ideal form of liberty for the bourgeois is 'freedom from; freedom from the constraints of obligations to wider society' (8). Yet, the bourgeois must remain socially engaged in order to secure the conditions needed for their own reproduction (ibid.). This social engagement is necessary because 'the bourgeois consider an empowered and politically conscious working class as "the incarnation of absolute evil", enemies of society that must be exterminated' (Engels, 1848, cited Ross, 2017: 81). Moretti explains that the bourgeois is by no means a uniform or stable group of actors. The bourgeois is the embodiment of capitalism, but not entirely congruent with capitalism. It is a community with very open borders – 'its very survival is dependent on new entrants with new ideas, innovative entrepreneurs who can refresh the lifeblood of capitalism and hence reproduce the social and economic conditions that give rise to the bourgeois' (Moretti, 2013: 4). During industrialisation in England, cities such as Manchester saw the rich living cheek-by-jowl with the propertyless classes. This friction created by the wide chasm between the two classes threatened the social order and a new "middle-class" was required to set an example for the poor, offer the illusion of the chance to climb out of poverty into the new "middle-class" and police the lower orders on behalf of the ruling class (Moretti, 2013: 11–12). Although the capitalist system that gave birth to them is one built on eternal change and destruction, another characteristic of the bourgeois is their role in promoting the goal of regularity, order, common sense and stability in human affairs (Ghosh, 2016). These features, combined with the fetishisation of efficiency, are manifest in the climate guardrails that corral climate discourse within the terms acceptable to bourgeois ideology.

0.4 Geographical focus

This book centres its analysis on the climate discourses of Europe and the West. This is not to marginalise other perspectives but – through recognition of the dangers posed by this capture of the discourse by a narrow set of perspectives – to open up the space for other worldviews to enter the discussion. The West is where our current understanding of climate change took root, and the West is still the locus of climate discourse and climate action. The European Union (EU) has commonly been recognised as a world leader in the building of international climate agreements (Zito, 2011). The climate crisis gestated in the heyday of liberal benevolence, in the womb of the European Enlightenment. Europe is the home, the mother of liberalism. It was here that liberalism enjoyed its greatest flowering, and from whence it has spread throughout the rest of the world. Though the liberal Enlightenment of the West envisaged that all human beings have equal moral status (Lukes, 2017: 14), de Tocqueville saw that in fact, 'we should almost say that the European is to other races what man himself is to the lower animals; he makes them subservient to his use, and what he cannot uses he destroys' (ibid). The focus on the West is not a rejection of the need for the decolonisation of knowledge, but is an effort to point out that much discussion of decolonisation of the research agenda or equalising of knowledge forms will be at best partial, because so much of that debate still take place within the guardrails of climate liberalism (Chapter 3). As Reiter writes in his introduction to an edited collection of essays on decolonisation of knowledge, 'we all need to rethink what development, growth, political power, democracy, nationalism and self rule mean and can mean – but the traditional Western approaches of European science do not contain the tools to ask different questions and find new and different answers' (2018: 2). Our focus on Europe and the West is intended to help us ask those different questions.

0.5 Why liberalism's time is up on climate change

The claim that liberalism is not up to the challenge of climate change begs the following questions:

a what is the evidence for the claim that liberalism cannot generate the policies needed for an effective response to climate change?
b what alternative to liberalism can meet the challenge of climate change?
c how is this alternative to be brought about?

Reflections on questions b and c are explored in the interviews with climate change researchers and communicators discussed in Chapter 7, though this book is intended to describe what actually is, rather than making claims about what ought to be. I will leave the search for happy tidy endings to the appropriate genres of fiction. As for question a., it is not possible to predict the future, but we can posit the theory that any response to climate change will be constrained by the terms set by the liberal framework of the West. By an "effective" response to climate change, I mean a set of policies and actions that can deliver the kind of future actually represented in liberal depictions of our net zero future. These depictions show the following:

- A world free from climate impacts.
- A world where capitalist economies are functioning and providing all the goods and food we want.
- A world where there is no social or class conflict.

If liberal climate policies cannot deliver this vision, then we should recognise those liberal depictions of our net zero future for what they are – ideology.

In terms of the evidence for the (in)ability of liberalism to provide the response it promises, liberalism has already failed on climate change. Over thirty years after climate change entered onto the world stage, atmospheric concentrations of greenhouse gases are accumulating at record-breaking speeds (NOAA, 2018), and according to some estimates, are now at their highest level for 15 million years (de la Vega et al., 2020). It is liberal capitalism that created the institutions and policy frameworks for managing that risk (Chapter 3), and it is those institutions and structures that failed to create the change needed 30 years ago that may have allowed us to create the future currently promised by existing net zero discourses.

The policy goal of net zero emissions by 2050 is the keystone that holds liberal climate hegemony in place. All manner of actions and technologies aimed at delivering net zero by 2050 are described as 'solutions'.

Animations, films and imagery of 2050 show a 1.5°C world free from climate impacts. This is as potent a form of climate science denial as you will find anywhere.[1] The science is clear – a 1.5°C warmer world is a very dangerous and unpleasant place and will remain so for the indefinite future (MacDougall et al., 2020). That is the most hopeful and optimistic outcome we can currently hope for, a new world lived permanently at the edge of catastrophe. Even this prospect is now seemingly out of reach, as the most ambitious international climate agreements – should they be implemented in full – still means 2A°C of warming, almost double the warming that is already wreaking havoc for life on earth (Meinshausen et al., 2022).

At the time of writing – summer 2022 – 1.1°C of warming is causing extreme weather events previously unobserved in the historical record, and these impacts are arriving much faster than climate science anticipated (Witze, 2022). The latest review of climate impacts from McKay et al. (2022), confirms the warnings from over 30 years ago (Rijsberman and Swart, 1990) and shows we left a safe climate behind at 1°C of warming. McKay et al. (2022) write that at current levels of warming, five climate tipping points (CTPs) are possible: Greenland and West Antarctic ice sheet collapse, tropic coral reef die-off, widespread abrupt permafrost thaw and Labrador Sea convection collapse. More importantly, at 1.5°C of warming (the goal of current international agreements), four of these CTPs become likely, and another five CTPs become possible (Atlantic Meridional Overturning Circulation collapse, Barents Sea ice collapse, mountain glaciers loss, boreal forest southern dieback / northern expansion). In total, the analysis from McKay et al. suggests up to six likely and four possible CTPs *below* 2°C. Even the most optimistic scenario for humanity is a world permanently 1.5°C warmer than the pre-industrial average and so a world forever subject to extreme and enduring climatic chaos. This outcome can only be described as a "solution" by glossing over uncertainties and known risks with wishful thinking, assuming technological salvation awaits just around the corner and pretending that Western social and economic systems – already buckling under at 1.1°C of warming – can continue operating as we are used to at 1.5°C of warming.

0.6   The structure of this book

Chapter 1 provides an in-depth analysis of the relationship between liberalism and the language of climate change, and a detailed explanation of the five guardrails. Each of the subsequent five chapters explores, one by one, the application of the guardrails in climate discourses. Treatment of these issues is complemented with data from analysis of 14 interviews with opinion-formers, decision-makers and campaigners, showing what models of human nature and our possible futures guide their approach to the politics of climate change governance (Chapter 7). These interviews show widespread disquiet at the limited options offered by working within the current system, but no clear sense of what the alternatives are and how we might get from "here" to "there", wherever "there" is. Our ideological satnav is broken, it has only one destination programmed into it and no one can work out how to put in a different destination.

The way liberalism has put the individual front and centre of climate action is discussed in Chapter 2. Chapter 2 goes on to examine the idea of the individual as agent and creator of her own future, by drawing on evidence from Marxist influenced theses about the relationship between the individual and society. Chapter 3 provides a genealogy of the institutionalisation of liberal norms into an international climate change governance regime. The universalisation of human nature based on a liberal ideal type shaped the philosophies within which a range of global institutions such as the UN arose. The capture of science by institutional liberal norms based on this universal view of the liberal human type is traced through the history of the language used by the IPCC and other key science-policy institutions. The evolution of climate change as a global scale phenomenon which can be managed through a universal temperature-based targets regime is critically examined as an instrument of discipline and control, rather than an empirical product. Chapter 4 discusses the construction of climate change as a crisis that does not necessitate the truly historical change that the IPCC identifies as necessary for a livable and flourishing future. Chapters 5 and 6 look at how this ahistorical approach is maintained. First, Chapter 5 proposes the idea that climate change is political all the way down, as opposed to the liberal position, namely that climate change is technical all the way down. Academic, campaign and policy narratives are examined to demonstrate the universality of the liberal faith in the power of technology to offer salvation. The ideas of technological determinist philosophers are contrasted with socially progressive left perspectives. These latter perspectives assume that industrial technology can be harnessed and controlled for the greater good, and the path of technological innovation is compatible with greater democracy and freedom, and that democracy and freedom are synonymous with a zero carbon capitalist economy.

In Chapter 6, it is argued that the turn to storytelling in the climate social sciences seeks to extend the liberal imaginary into the heart and soul of every climate concerned citizen. The "narrative turn", "new stories will save us" and the need to find "new myths for the 21st century" – these ideas increasingly dominate discussions of how to build public support for ambitious climate change policy. Such approaches are discussed as an effort to provide scenarios which offer an alternative to social conflict and historical change.

Following the interview analysis and discussion in Chapter 7, Chapter 8 summarises the arguments offered in the book and touches on options for a non-liberal response to climate change. Instrumental policies and discourses of controlling nature are contrasted against more metaphysical, deep green, non- rational beliefs. Chapter 8 will argue that liberalism's pyrrhic victory is assured, not because it is the best political philosophy, but through the adept way it has co-opted climate change.

At the time of writing liberalism's victory over the debate about how we should respond to climate change feels complete and permanent.

0.7   Conclusion

Liberal norms encode the way in which our culture represents climate change, and police what it is possible to think and say about climate change. Understanding this limitation and breaking through that to a future of alternative possibilities is particularly important at a time when more people are becoming aware of climate change, some of whom will be exploring in more depth what our future holds, and what responses are being discussed as acceptable and reasonable. However hopeless the situation may appear, it seems appropriate to try and understand what, if any relationship, exists between liberalism and climate change, and what sort of impact that relationship may be having on our failure to do anything meaningful about climate change.

Note

1   A Google image search using the term 'net zero future' will illustrate these three points.

References

Anderson, P. (2017). *The H-Word. The Peripeteia of Hegemony.* London: Verso.

Bradshaw, C., et al. (2021). 'Underestimating the challenges of avoiding a ghastly future'. *Frontiers in Conservation Science.* doi: 10.3389/fcosc.2020.615419.

Ceballos, G., Ehrich, P. and Dirzo, R. (2017). 'Biological annihilation via the ongoing sixth mass extinction signalled by vertebrate population losses and declines'. *PNAS,* Vol. 114, 30.

Davis, S., et al. (2018). 'Net-zero emissions energy systems'. *Science,* Vol. 360, No. 6396. https://doi.org/10.1126/science.aas9793

de la Vega, E., et al. (2020). 'Atmospheric CO2 during the Mid-Piacenzian Warm Period and the M2 glaciation'. *Science Reports,* Vol. 10, 11002.

Fisher, M. (2009). *Capitalist Realism. Is There No Alternative?* Winchester: Zero Books.

Geuss, R. (2022). *Not Thinking Like a Liberal.* London: Belknap Press.

Ghosh, A. (2016). *The Great Derangement – Climate Change and the Unthinkable.* Chicago: Chicago University Press.

Hobsbawm, E. (1988). *The Age of Capital.* London: Abacus.

IPCC. 2021. Climate change widespread, rapid, and intensifying. Retrieved from https://www.ipcc.ch/2021/08/09/ar6-wg1-20210809-pr/.

IPCC. (2022). *Overarching Frequently Asked Questions and Answers*. Retrieved from https://www.ipcc.ch/report/ar6/wg2/about/frequently-asked-questions/keyfaq6/.

Knight, B. (2019). *Amitav Ghosh: What the West Doesn't Get About the Climate Crisis*. DW. Retrieved from https://www.dw.com/en/amitav-ghosh-what-the-west-doesnt-get-about-the-climate-crisis/a-50823088. Accessed 27/11/2021.

Lukes, S. (2017). *Liberals and Cannibals. The Implications of Diversity*. London: Verso.

MacAskill, W. (2022). *The Beginning of History. Surviving Catastrophic Risk.* Foreign Policy. Retrieved from https://www.foreignaffairs.com/world/william-macaskill-beginning-history.

MacDougall, A. H., et al. (2020). 'Is there warming in the pipeline? A multi-model analysis of the Zero Emissions Commitment from CO2'. *Biogeoscience*s, Vol. 17, 2987–3016.

Mann, G. and Wainwright, J. (2018). *Climate Leviathan*. London: Verso.

Meinshausen, M., et al. (2022). 'Realization of Paris Agreement pledges may limit warming just below 2 °C'. *Nature*, Vol. 604, 304–309. doi: 10.1038/s41586-022-04553-z).

McKay, D., et al. (2022). Exceeding 1.5°C global warming could trigger multiple climate tipping points. *Nature*, Vol. 377, No. 6611. doi: 10.1126/science.abn7950.

Moretti, F. (2013). *The Bourgeois*. London: Verso.

Moscovici, S. (1963). Attitudes and opinions. *Annual Review of Psychology*, Vol. 14, 231–260.

NOAA. (2018). NOAA's greenhouse gas index up 41 percent since 1990. Retrieved from https://research.noaa.gov/article/ArtMID/587/ArticleID/2359/NOAA%E2%80%99s-greenhouse-gas-index-up-41-percent-since-1990.

Porter, G. and Brown, J. (2001). *Global Environmental Politics*. London: Routledge.

Reiter, B. (2018). *Constructing the Pluriverse: The Geopolitics of Knowledge*. London: Duke University Press.

Rijsberman, F.R. and Swart, R.J. (Eds.) (1990). *Targets and Indicators of Climatic Change*. Stockholm: Stockholm Environment Institute.

Ross, A. (1991). S*trange Weather.* London: Verso.

Ross, K. (2017). *Communal Luxury. The Political Imaginary of the Paris Commune*. London: Verso.

Shah – Kazemi, R. (2006). *Paths to Transcendence*. Indiana: World Wisdom.

Steffen, W., et al. (2015). 'Planetary boundaries: Guiding human development on a changing planet' *Science,* Vol. 347, No. 6223. doi: 10.1126/science.1259855.

Streeck, W. (2016). 'Heller, Schmit and the Euro', in *How Will Capitalism End*. London: Verso, pp. 151–164.

UNFCCC. (2017). *International Agreements Must Drive Deep Transformation of Societies*. Retrieved from https://unfccc.int/news/international-agreements-must-drive-deep-transformation-of-societies.

WBGU – German Advisory Council on Global Change. (1995). *Scenario for the Derivation of Global CO2-Reduction Targets and Implementation Strategies*. Special Report. Berlin: WBGU.

WBGU – German Advisory Council on Global Change. (1997). *Targets for Climate Protection 1997*. Special Report. Berlin: WBGU.

WBGU – German Advisory Council on Global Change. (2003). *Climate Protection Strategies for the 21st Century. Kyoto and Beyond*. Special Report. Berlin: WBGU.

WBGU – German Advisory Council on Global Change. (2009). *Solving the Climate Dilemma: The Budget Approach*. Special Report. Berlin: WBGU.

WBGU – German Advisory Council on Global Change. (2014). *Human Progress within Planetary Guardrails. A Contribution to the SDG Debate*. Policy Paper No. 8. Berlin: WBGU.

Whyte, H. (1973). *Metahistory. The Historical Imagination in Nineteenth Century Europe.* Baltimore, MD: John Hopkins University Press.

Witze, A. (2022). 'Extreme heatwaves: Surprising lessons from the record warmth'. *Nature*, Vol. 608, 464–465. doi: 10.1038/d41586-022-02114-y.

Zito, A. (2011). 'The European Union as an environmental leader in a global environment'. *Globalizations,* Vol. 2, Vol. 3, 363–375.

1 The struggles of climate liberalism

1.1 Sublimating paradox

'Liberalism is in trouble' exclaimed a headline in the Wall Street Journal in 2021 (Swaim, 2021). This lament has been a perennial concern of liberalism since the early 20th Century (Grief, 2015). The characteristics of liberalism described as being under threat include, but are not limited to, formal representative democracy, an embrace of free markets, limits on state power, the separation of state and church and individual freedom. Modern threats to liberalism are to be found either within liberal societies themselves (e.g. in the United States from Donald Trump and his supporters) or from other non-liberal societies (Russia, China, Iran, Libya, Iraq, Vietnam, Korea etc.).

There is no universally agreed scale for comparing crises. That said, the very irreversibility of climate change tipping points – which threaten the existence of civilisation (Ehrlich and Ehrlich, 2013) – sets it apart from economic or social crises. Banks can pump money into the system to keep markets afloat. Politicians could commit to creating a more equal and inclusive society with more opportunities for participatory deliberation. Nations could disarm. Despite the enormity of climate change tipping point risks, climate change is not widely recognised as a threat to liberalism in mainstream commentary of the health of liberalism. For some climate change scholars, the limits of "formal" democracy (e.g. a vote every four or five years) is seen as a problem and it is argued that an effective climate response requires more democracy (Fischer, 2017; Machin, 2013). Mann and Wainwright (2017) and Fischer (2017) are some of the authors raising the concern that authoritarian states may emerge in response to increasing climate impacts, jettisoning what little democracy remains. The role of these authoritarian states will be to protect the interests of the rich from climate impacts and an increasingly desperate and unruly public.

One feature of liberalism is the desire to treat social problems as separate from each other. Rising inequality is not, in this view, a problem of liberalism, a systemic issue but is instead a problem with how people are selected for jobs, and the answer is to implement measures to promote targets to improve meritocracy in recruitment practices. This desire to treat problems as separate from each other is manifested in liberal climate policy which seeks to reduce emissions within the framework

DOI: 10.4324/9780429463488-2

of liberal free market policies, rather than see climate change as a problem that cannot be addressed within liberal free market policies. Yet, increasingly climate change is breaking down the distinction between public and private, which is one of the hallmarks of liberalism. My private actions as a consumer have impacts on the public sphere and my private actions can no longer be separated from the public sphere. The private has implications for the public. Perhaps it always has but now that tension is made stark with climate change.

Liberalism contains many contradictions and unresolved tensions, but it cannot face up to or live with paradox. Instead, it rules by reason, logic and contradiction-free argument (Groys, 2014: 21; Stehr and Grundmann, 2012). Where there is paradox, it is either simply ignored or managed through ideology, made not to seem a paradox. For example, in climate discourses 2°C or 1.5°C of warming is sometimes called a "safe" limit, sometimes a "dangerous" limit or the idea of a limit is dismissed as arbitrary (McGuire, 2022), but then still continues to be used as a defining principle of a purportedly scientific response to climate change. The UK Business, Energy and Industrial Strategy (BEIS) has commissioned an independent review of the government's approach to delivering its net zero target, to ensure that net zero is being delivered in a way that is 'pro-business and pro-growth' (BEIS, 2022). The idea that the goals of ending anthropogenic forcing of the conflict and the pursuit of economic growth may be in conflict with each other, may present a paradox, is absent from this call for evidence.

Climate change cannot, despite efforts to reduce the discussion to number and money, be controlled within the narratives that seek to deny these contradictions. The limits of liberalism are exposed by climate change, and this failure is expressed in a liberal climate discourse which is 'confused, contradictory and chaotic' (Ereaut and Segnit, 2007). The difficulty , as Anderson writes, is that liberalism does not result in a unique set of principles (Anderson, 1987: 342). It is possible to order the guiding values of liberalism in a variety of ways and the meaning of such fundamental concepts as right to freedom and equality will always be contestable (ibid.). When it comes to climate change, uncertainties, ambiguity and contradiction are helpful to liberalism's strategies for accommodating paradox. For Anderson, 'A hegemonic discourse is never statistical, but performative. For such a performance specifics can be an impediment, vagueness a virtue' (Anderson, 2017: 97). Anderson provides another example of this paradox when he writes that lying is wrong, but there may be circumstances in which lying seems a moral imperative (ibid.).

The question 'would one uphold the ideal of free market choice in the face of irreversible environmental deterioration' (Anderson, 1987: 28) has a clear answer – yes. The paradox is thus solved. In this instance at least, liberalism's moral scale is free of ambiguity and confusion – free market choice is more important than the maintenance of a viable biosphere. In a similarly optimistic tenor, de Zeongotita offers a brave defence of what liberalism *could* be.

What we can say is that liberalism means accepting a world system of private enterprise, and technological innovation and consumer culture, and you want to see it managed so that no one is excluded and the environment is protected, free expression flourishes and so on.

(de Zengotita, 2003)

Twenty years later, the wait for environmental protection goes on. Perhaps it is time to accept that liberalism and a liveable climate are irreconcilable.

Climate change is, like a receding glacier, forcing liberalism to give up its secrets, bringing many of the contradictions and paradoxes at the heart of liberalism into the light of day. Not the least of these growing tensions is that which emerges when trying to square the circle of climate science with the ideology of liberal freedoms. Peckham wrote that 'an ideology is always out of phase with the situation in which it is employed, for an ideology has always emerged as a response to a preceding situation' (Peckham, 1995: 12). Hence whilst the ideological stories of liberalism accompanied and justified the imperialism of the West, the break from religious persecution and feudalism, and propelled the potential of scientific innovation, those same liberal ideologies are not geared up for climate change. So it is that Peckham concludes that 'undermining the ideological superstructure of Western culture is humans only hope' (1995: 14).

As climate change becomes increasingly co-opted by liberal culture, the possibility of looking beyond the boundaries set by liberal climate "guardrails" becomes ever more remote. In his despairing search for an alternative to 'capitalist realism', Fisher explains that our inability to even think it possible to look beyond liberalism for an answer to climate change is a sign of ideology in action. Ideology succeeds when it is fully naturalised, when we can no longer recognise it as ideology (Fisher, 2009: 16). Citing Zupanci, Fisher records that the highest form of ideology is the ideology that represents itself as empirical fact (ibid: 17). The Western definition of a scientifically derived single limit for the world is the most prescient example of ideology as fact. The reduction of climate change to net zero policies becomes entrenched as it spreads and is repeated throughout Western culture. The contradictions, the ambiguities and the risks of liberal climate policies are hidden in these simplistic narratives of net zero solutions. What looks like the answer is just a deepening of the crisis. Given that liberalism itself has been in crisis since its inception, and has been the progenitor of ever-deepening catastrophe since the beginning of the 20th century, this should be no surprise.

1.2   The best of all possible worlds, the worst of all possible worlds

Liberalism has no answers for climate change, other than to aim for a "least worst" outcome, whilst staying true to the "least worst" political philosophy.[1] Liberalism rejects the search for an ideal utopia as naive, and offers in its place a liberal market utopia (Whyte, 1973: 38). This "least worst" utopia is the latest instalment of civilisation's search for unity between the individual and the collective. What makes liberalism different from other efforts to make whole that which civilisation has ripped apart, is that it gives primacy to the individual over the collective. The prioritisation of the individual over the collective has progressed to the point that the collective or communal is reduced to an afterthought, except in so much as the search for the free expression of one's individual will is the one thing that unites humanity, the one thing we all have in common. Exactly when our fall from the state of grace of perfect unity happened is debated – the advent of agriculture is a common marker, though others trace the schism back further, to the advent of symbolic activity[2] whilst for others it is not agriculture but the emergence of the state that signalled a life of alienation (Scott, 2017). Hegel argued the modern state was an expression of the long sought-after unity, though Marx thought that class divisions negated any claim of unity deliverable by the state. Resurrecting a Hegelian vision, Fukuyama's end of history thesis (Fukuyama, 1992) suggests the residual friction between the universal and particular will remain, and that we will progress no nearer to that unity than we have already reached under liberal democracy.

Conservatism is the defence of an already existing perceived unity, expressed through a stratified society reflecting the natural order in which everyone knows their place and stays in it. Conservatism is not the defence of an ideal past but of the present social dispensation (Whyte, 1973: 22). Conservatives pretend to value individualism, but actually defend the status quo by showing it to be the integrated, organic unity that anarchists can only dream of achieving (ibid: 23). Such conflict as occurs between the centre-left and centre-right is largely limited to cultural issues. The kind of bourgeois liberalism being discussed in this book is pro-market on economic issues and anti-traditional on social issues, whereas the working class are more to the left on economic issues and conservative on social issues (Lind, 2020: 73). Hence for both conservatives and liberals, there are common interests on the economy, and it is the cultural arenas that offers a safe space in which the centre-right and centre-left can fight out what small differences exists between the two philosophies. As North highlighted in his review of the politics of literary criticism, liberalism and conservatism are but two wings of liberalism (North, 2017: 8). This seems especially the case for climate policy.

Tocqueville saw that the prospect of revolutionary change was unlikely in liberal democracies because it is the most effective political system for delivering material security. As people become more materialistic, they will have more to lose from radical change and so will become more reactionary in their politics ((1835) (1983)). All are catered for by the consumerist society, so none may escape (Ellul, 1973), or wish to escape, even if they could think of where to escape to. Baumann writes of consumerism that it has become 'the cognitive and moral focus of life' (1992: 49), the 'playground for individual freedom' (ibid: 51) that has the 'crucial task of soliciting behaviour functionally indispensable for the capitalist system and at the same time harmless to the capitalist system' (ibid: 52). What I want to suggest here is that rather than being a means by which we differentiate ourselves from others through building an identity (as suggested by, amongst many others, Lodziak, 2000: 114), consumerism is an easy way to rebuild connections with others by using our purchases to identify ourselves as part of a larger community with shared interests. We are using consumption to repair the ties broken by liberalism.

In sum, the mainstreams of left and right politics are both dedicated to the maintenance of the systems that provide and protect material security for a sufficient number to secure political stability. Its definition of social being, based on a Hobbesian fear of the Other, reinforces a commitment to Lockean private property for the Self, and invariably leads to policies of annihilation/conversion (Agathangelou and Ling, 2006).

1.3   Freedom from, or freedom to?

Freedom is ultimately quite a meaningless word, and in liberal societies is utterly dependent upon a legal framework for its implementation, a framework which denies some people freedom. It is possible for "free" societies to function in the throes of such a contradiction precisely because freedom is a confused and arbitrary concept with little practical or social function (Houllebecq, 2000, cited Betty (2017: 88). Liberalism refuses to recognise these contradictions, except at the most superficial level (Rodney, 2018: 12). This contradiction is manifest in the response of liberal politics to climate change. We are free individuals, free to choose the future we want. Except, in respect of the most important issue facing humanity, climate change, we are not free to choose the future we want. We are all obliged to live with the risks of at least 1.5°C of warming.

Liberal ideas of freedom claim that humans can become free in spite of, not through, social relations (Caudwell, 2018). Claims that one's freedom and happiness can be found only through individual effort are based on an illusion, and the desire to maintain that illusion is one of the drivers of liberalism's ceaseless crisis. 'Liberals, the idealists, are all seeking the impossible solution, salvation through the free act of the individual will amid decay and disaster' (Caudwell, 2018: 21).

Sometimes the source of our woes is blamed not on liberalism, so much as a particular form of liberalism, namely neoliberalism. As Zizek points out, it is better not to criticise the system per se, but rather point to an idealised version of the system that we might one day get (back) to (Zizek, 2011). For holders of this view, neoliberalism, in freeing capital from any sense of social obligation through, for example, the use of taxes to redistribute wealth or regulation to limit environmental harm, has undermined the social cohesion that classical liberalism was able to maintain, and which is necessary for sustainable living. Hedges claims the classical liberalism of the late 19th century saw a growth of mass movements and social reforms that addressed working conditions in factories, the organising of labour unions, womens' rights, universal education and housing for the poor, a golden era of liberalism that ended with the first world war (2010: 7). This, for Hedges, was liberalism performing its proper role of acting as a safety valve for democracies by overseeing a slow programme of reform that offers people hope (Hedges, 2010: 8). Today the liberal agenda has – in the form of neoliberalism – been captured by corporate interests, reducing the liberal class to 'a useless and despised appendage of corporate power' who were once the voice of resistance to an excess of profit and greed, but who are now reduced to producing the cultural outputs that keep people trapped in illusions (ibid: 13). Hedges is moved to describe this situation as 'tragic', because the liberal class no longer contains the ability to reform itself, and the power void it has left behind has been filled by the interests of the right. This is not so much a defeat of classical liberalism but a complete inversion of liberalism's original goals (ibid.: 15).

In *Crisis and Sustainability: The Delusion of Free Markets* (2017), Alessandro Vercelli charts the rise and fall of two different definitions of freedom in liberal thought. "Negative freedom", describing an agent's freedom from specific constraints imposed on his (sic) potential or actual actions, and "positive freedom", defined as 'freedom to act for realisation of the agent's goals' (Vercelli, 2017: 33, emphasis in original). The author notes that liberalism, like freedom, remains a contested term, but that all definitions are predicated on the assumption that freedom is good and any limit on freedom must be thoroughly justified (ibid.: 40). Negative freedom, imagined as freedom from interference by the state, is the political philosophy which underpins the neoliberal economic agenda. The author argues that neoliberals have falsely maintained that the negative and positive forms of freedom are mutually exclusive, whereas in fact, they are two 'complementary aspects of liberty, not irreconcilable conceptions of it' (ibid: 34). Vercelli argues that the solution is for the state to have a greater degree of control over the economy, and to direct economic activity towards the building of a financially, socially and environmentally sustainable future: 'we cannot dream of being able to converge towards a sustainable development trajectory without abandoning the neoliberal point of view and its policy strategy' (264). The initial period of classical liberalism began in earnest in 17th-century England and was given theoretical confirmation by Adam Smith in The Wealth of Nations. Vercelli argues that the perfect market conditions under which Smith's famous 'invisible hand' operates do not exist (40). Markets are not perfectly competitive. Neoliberalism is the idea that the answer to this failure is to further liberalise the conditions under which markets operate. Following the "Great Depression" of the 1930s, there emerged a consensus that laissez-faire economic policies justified by a negative concept of freedom were creating evil in the world, and a positive definition of freedom with the state intervening to ensure markets were a force for good (classical liberalism) came to dominate economic policy until the 1970s. Vercelli longs for a return to a classic liberalism as the basis for a sustainable future, regardless of the war on nature that characterised this supposed heyday of liberalism. Whether such a thing as neoliberalism even exists is disputed, no more so than by those to whom the label is attached (Mirowski, 2014: 39). Given that accumulating wealth over long stretches of time requires a strong state that can back the laws that protect the wealthy (Pistor, 2019: 1), we are best served, for the purposes of the analysis in this book, in disregarding claims that the state is less present in the liberalism of today than 50 years ago.

For Foucault, freedom expresses itself through practice; it only exists when it is exercised and it is exercised in resistance to power (Foucault, 2007, cited Cook, 2018: 104). Or as Luxemburg said, we only notice our chains when we try to move. The effort to understand how power operates through language to constrain our possible responses to climate change within liberal frameworks is an attempt to exercise this Foucauldian definition of freedom.

1.4   Anarchy and order

The anarchy of the free market has to somehow co-exist with the bourgeois desire for order. The "guardrails" examined in this book are intended to impose an order on climate change policy whilst leaving the anarchy of the market intact. The desire for order has always been in conflict with the individual liberalism of the marketplace, and poses the question, how can individualistic societies respond to climate change if the individual pursuit of happiness creates disorder in the climate and consequently disorder in societies dependent on a stable and predictable climate?

Some commentators have identified consumer society itself as a threat to the liberal order. Democracy is dying because consumerism and culture can fulfil human needs more directly and efficiently than politics ever could. From this perspective, the reduction of existence to the nihilistic blind pursuit of desire is undermining civilisation and ushering in a new era of barbarism (e.g. Grief, 2015; Lukes, 2017). Liberal democracies have thrown open the doors to a previously exclusive party. Everyone is welcome, as long as they don't spoil the vibe or piss in the punch bowl. In this world, nothing is better than anything else. There is no goal to progress but progress itself, the only measure of value is the amount of pleasure we can extract from the 'everything, everywhere, all the time' anarchic marketplace of modernity (Jeffries, 2021).

Spufford, in his book "*Red Plenty*", traces the Soviet Union's defeat in the Cold War to its efforts to centrally plan how to meet people's desires. He argues that people's needs can be better met through the anarchy of the market than centralised rational planning. A single scientifically designed programme for happiness will never be able to meet people's needs and desires as effectively as simply allowing each individual to pursue their own irrational idea of happiness through the markets (Spufford, 2010: 269). For defenders of free market liberalism such as Spufford, all meaning and happiness is derived from consumption, and the best system is the one that can meet this ceaseless need. The only trouble with this claim is the large body of research showing that once basic material needs are met, most people find the satisfaction derived from time with friends, family and living as part of a community, far outweigh the pleasures of consuming superfluous commodities (Gorz, 1999).

This abhorrence of centralised planning is a defining feature of liberalism. It is a well-founded fear, and Elinor Ostrom was surely correct to argue that planners whose lives are distant from the lives of those they would govern do not have the knowledge of those lives needed to govern effectively (Wall, 2014). And yet climate policy seeks to centrally plan, at a global scale, the response to climate change through the claim that there is a single knowable dangerous limit to climate change. This meta target is to be met by using elements of market mechanisms, keeping individualism the primary operating principle, the ultimate ontology. It is just such unresolved tension, contradiction and confusion which lies at the heart of our failure to respond effectively to climate change.

The anarchy of the marketplace has to be tightly bounded, and managed through a strong centralised state (Pistor, 2019) to provide stability and some social order such as is needed for the running of a 'technological society' (Ellul, 1965). Religion, in the early days of liberalism, was meant to offer order and stability in an otherwise rapidly changing world. Nadler's history of this period records the hope that a shared official religion, such as that of Protestantism, would weld rulers and subjects together under the divine protection that depended on an orderly religious life regulated by true doctrine, a well-ordered church organisation, decent public worship and pious public conduct. But a careful balance had to be struck to ensure this new state-sanctioned religion did not become so powerful as to pose a threat to the secular and commercial affairs of the nation (Nadler, 2011: 26). That balancing act continues today, but with the law, rather than the threat of eternal damnation, now doing much of the work needed to protect the secular and commercial affairs of the West. Pistor (2019) shows how the law is 'a powerful tool for social ordering... which has been placed firmly in the service of capital' (Pistor, 2019: xi). The idea that individual initiative protected by clear property rights would ensure scarce resources would be allocated to the most efficient owner, increasing everyone's wealth, contains a degree of logic. But, writes Pistor, the capture of these laws by special interests has led to levels of inequality not seen since before the French Revolution (ibid, 2019: 2).

In any large modern society, one would expect to encounter a plurality of values amongst the population. The liberal state deals with these differences of opinion, whilst still providing some democratic choice and social stability, in part by removing large areas of decision-making from the political sphere or substituting value choices with economic calculations. We should not expect to find universal ideas of what is good and right amongst the citizens of a free, open and liberal society. Nor should we expect that any definition of acceptable risk is going to be the same for everyone. Yet whilst liberalism claims the individual and their values are paramount, the individual has no choice in respect of climate change, but to accept the single measure of acceptable risk imposed on everyone by politicians.

1.5   Openness to new ideas vs the reproduction of liberalism

Groys defines liberalism in relation to communism, arguing if communism is the project of subordinating the economy to politics, in order to allow politics to operate sovereignly, then liberalism is the project which seeks to allow politics and the economy to operate with equal power (2014: 20). But the economy operates in the language of number, in the language of money, whereas politics operates in the language of values and thought (ibid.). One cannot enter into discussion with economic processes, all one can do is adapt to the demands of money. This requirement to speak of political, moral and value issues in the language of money, number and efficiency reduces what it is possible to imagine and dream for in terms of what already exists. It is a framework that makes any desire to abandon this liberal utopia appear childish, irrational and naive. Instead, we judge the value of our actions as though making entries into a ledger, as if life cannot be anything other than that whose worth can be assessed through double-entry book-keeping. What is the profit and what is the loss of this plan, what are the costs, and what are the benefits? (Shaw and Nerlich, 2015). 'We need to be realistic, get things done, and accept that getting on with one's life almost requires that a person become an enemy of both humanity and a healthy biosphere' (Winner, 1986: 83).

Spark has written a history of the afterlife of Herman Melville's Moby Dick, detailing how it was received and discussed in the US around the time of its publication. She describes how, in the lead up to the publication of Moby Dick, the social and economic upheavals that accompanied the Enlightenment led to great anxiety amongst political and financial elites, as the old agrarian order broke down, and people, uprooted from tradition, custom and norms, began leaving the countryside and flooding into the cities, where the powerful and rich people lived. Much political, cultural and social history of the last 200 years is best explained as an effort on behalf of those elites to control the people whilst convincing them they were free. Enlightenment rationalism and universal standards of morality were enlisted in the service of amelioration and democratic self-management, ener-gising resistance to all forms of radical politics (Spark, 2001). Through govern-ment regulation of markets and industry, and a partial redistribution of wealth and resources, alongside incorporation of the dissonant elements (labour unions and minority groups), progressives sought to reinstate the equilibria thought to be lost during the transition from a peaceful and integrated agrarian society to the urban and industrial nightmare of market-induced greed, exploitation, economic strife and anomie (Spark, 2001: 11). These experts and planners gathered facts, professed scientific objectivity but they were also paternalists attempting to preserve, recover or retain class power and authority (ibid.: 12). In his examination of how salaried professionals learn to play the game, Jeff Schmidt recounts how it was the most well educated who can be most trusted to act in the name of this bigger vision. The route to advancement and employment security is the ability to act with extreme political caution in all you say and do (Schmidt, 2001). The HR department knows this as well as the critical sociologist.

Roelofs (2003) identifies philanthropic actors as key players in the rigging of the market place of ideas. Roelofs – referencing Marx and Engels Communist Manifesto – points to the fact that the bourgeois seek to redress some of the prob-lems created by bourgeois society simply in order to secure the conditions for the ongoing reproduction of bourgeois society. The intellectuals are key to this goal, for political systems are most secure when this class of people are given interesting and well-rewarded work, work which reproduces hegemonic control by creatively incorporating emerging trends and challenges. Challenges such as climate change.

The framing of our possibilities in economic terms and the capture of public discourse by actors invested in maintenance of the existing order go someway to explaining how the promise of liberalism has ossified into a sclerotic condition in which the reproduction of liberalism is the only goal. We are powerless, una-ble to try new ways of organising our economy, destined to continue trying what we have always done, turn to the markets in order to trade our way to survival

(White, 2013: 91). For the middle classes, the market is a mirror of the timeless and immutable laws of nature, a sacred condition which must remain outside the flux of history (Noble, 2012: 8).

This ideological paralysis is cutting off our escape route out of climate catastrophe. The anaemic liberal discourses of climate risk are being circulated and amplified (and hence naturalised) by centre-left progressive voices that are presented as the champions of ambitious action on climate change. The centre-left gatekeeping of the discourse has been so effective that the ideological water in which we swim is no longer a part of the discussion. The only critique available is from the climate denying right wing.

Enlightenment liberalism was intended to provide the structures and processes that would allow for the fair deliberation of what counts as the good life. Societies progress through experiments in living, and thus liberalism should be an outlook, an attitude, not a fixed and immutable set of conclusions. A "free" society cannot be based on any particular creed, for example, liberalism. A free society is a society in which all traditions are given equal rights. What would define enlightenment liberalism is 'moral and political argument that takes as little for granted as possible and does not assume the value of any particular way of life or the justice of any particular set of social and political arrangements' (Lukes, 2017: 7).

Having discussed how climate change has exposed the contradictions and conflicts of liberalism, we now provide a fuller account of how those tensions are managed in climate discourses in the form of the five climate guardrails introduced earlier in this book.

1.6 The five liberal climate guardrails

Denial of climate change is on the wane (Black, 2018) and is now largely limited to comments on the discussion boards of right-wing websites. What is of interest to us is how notionally progressive voices coalesce around a narrative that accepts the enormity of climate change, but presents a reformist picture which substitutes that urgency for a story that sees the possibility for change limited to a quest for a different way of fuelling business as usual.

Picture a road, taking drivers from place to place, from here to there. One does not need to assume a shared ideology amongst the people taking that journey, anymore than one needs to assume researchers and campaigners, whose individuality does not extend beyond being nodes in the circulation of capital, are necessarily consciously promoting a particular ideology over others. The route has been defined by nameless people, as part of processes which are part of a larger, if only vaguely defined, vision. So, it is with the concept of the guardrails which steer the mainstream climate discourse. The ultimate goal is not very clear, other than continued economic growth. Nameless people, dispensing research grants, managing philanthropic funding, designing campaigns, are all working to this vaguely defined end, a fuzzy picture of a prosperous world of trading and production, free of climate change. These discourse domains, as well as journalism and politics are

all populated by representatives of a particular background. Anyone with a different background can enter into this managerial caste, as long as they adhere to the
norms of that institution, and can read the directions and road signs. In the case of
climate change, the important thing is the ability and willingness to act as a carrier
of the 1.5°C/net zero discourse.

A temperature guardrail describes a level of warming that is an upper limit to
be observed if humanity is to avoid dangerous climate change. An early use of
the concept described a guardrail approach as one that 'searches for cost-effective
mitigation that are designed to keep the future from venturing too close to evolving definitions of the boundaries of truly dangerous climate change along multiple dimensions' (Yohe and Toth, 2000: 104). A report of the proceedings from an
international science congress organised by the University of Copenhagen in the
lead up to the United Nations Framework Convention on Climate Change held in
Copenhagen in 2009 warned 'There is a looming biodiversity catastrophe if global
mean temperature rises above the 2°C guardrail' (Richardson et al., 2009: 13). The
German Advisory Council, leading architects of the 2°C policy framework (Jaeger
and Jaeger, 2010) have repeatedly referred to 2°C as a "guardrail" (WBGU, 1995,
1997, 2003, 2009).

Our repurposing of this concept examines linguistic, discursive and symbolic
guardrails that operate through language to define what is deemed a reasonable
and acceptable response to climate change. In essence, what is a politically safe
limit to the amount of social, economic, political and ideological change that can
be discussed as providing a response to climate change. The guardrails metaphor
for climate policy discourses is equally applicable to 1.5°C as it is 2°C, as in both
cases it operates to the same ends, to substitute policy with a misleading sense of
scientific certainty, which in turn, provides a rationale for an approach which is less
about climate change and more about liberal ideology.

Climate Guardrail 1: *Climate change is not a challenge to individualism.*

I am free to choose whether or not I engage in carbon intensive activities.
You are free to choose whether or not you engage in carbon intensive
activities. We are free to choose whether or not we engage in carbon intensive activities. Ultimately it is through our individual, rational free choices
that climate change is solved. The state and corporations will respond to
the signals they receive from individual choices, and enable the change that
individuals demand.

Climate Guardrail 2: *The liberal construction of climate change is
universally true.*

Western representations of climate risk and acceptable levels of harm sit
above ideology, national interests, culture and geography. There is one dangerous limit to climate change. It is the same whoever and wherever you
are. Using this target as a policy objective helps coordinate global responses
under the one right way of being in the world.

Climate Guardrail 3: *Climate change is not an historical phenomenon.*

Historical change and social conflict are irrelevant to solving climate change. Climate change is not a driver of historical change in the way that material forces have driven historical epochs in the past, from primitive communalism, through to feudalism and capitalism. Climate change can be solved within the existing relations of production that characterise liberal democracies. Climate change should be de-politicised, and responded to purely on the basis of the science and agreed targets.

Climate Guardrail 4: *We have the technologies to solve climate change.*

Human progress is real, is expressed through technological change and is an unassailable imperative of human existence. The alternative to this position is to plead that we go back to living in caves. We can solve climate change and keep the lights burning.

Climate Guardrail 5: *New stories will save us.*

We need new stories that can help humanity live sustainably. These stories will show how the whole of humanity can align with the norms of western liberal democracies, and live clean, smart, prosperous and fulfilling lifestyles without dangerous emissions of greenhouse gases. These stories will make the transition to a better future possible without the need for social conflict.

The following five chapters delve into more detail on each guardrail and how language is used to constrain the climate discourse within these five principles.

1.7   Conclusion

The paradoxes, contradictions and tensions of liberalism make the justification of liberal climate policies particularly difficult. This tension is most stark in the desire to "solve" climate change whilst continuing to create economic growth. This contradiction is masked by an idea which itself is in direct contradiction to liberal principles, namely the creation of a single dangerous limit for climate change in a world made up of diverse individuals with diverse needs and desires, all of which must be respected in the name of individual rights. A quick dive into the liberal language of climate change soon reveals many other tensions. Liberalism is the separation of social functions and social problems, each to be treated in its own terms. Church is separated from state, economics is separated from politics, private interests are separated from the public interest, we are separated from each other. Climate change, as well as being separated from other environmental problems, is also treated as simply a problem of emissions, not a systemic problem demanding rapid and revolutionary change to everything. Unlike other authors who might argue that this crisis 'changes everything' whilst not seemingly wishing to change much at all, this book argues that our response to climate change really should change much more than anyone is really willing to accept, especially the lucky citizens of wealthy Western liberal societies.

Notes

1 This is an allusion to Winston Churchill's famous quote 'democracy is the worst form of Government except for all those other forms that have been tried from time to time…'.
2 I am here referring to the work of green anarchists (most notably John Zerzan, cited as an influence by the UnaBomber), and some anthropologists (for example, Marshall Salins and Patrick Sale). For Marx, humanity's need to hunt and roam remains undimmed, if largely denied in modern Western societies.

References

Agathangelou, A. and Ling, J. (2006). *Transforming World Politics. From Empire to Multiple Worlds*. London: Routledge.
Anderson, C.W. (1987). 'The human sciences and the liberal polity in rhetorical relationship', in Nelson, J., Megill, A. and McClosekey, D. (eds). *The Rhetoric of the Human Sciences*. Wisconsin: University of Wisconsin Press, pp. 341–362.
Anderson, P. (2017). *The H-Word. The Peripeteia of Hegemony*. London: Verso.
Baumann, Z (1992). *Imitations of Post-Modernity*. London: Routledge.
BEIS. (2022). *Net Zero Review: Call for Evidence*. Retrieved from https://www.gov.uk/government/consultations/review-of-net-zero-call-for-evidence/net-zero-review-call-for-evidence
Betty, L. (2017). *Without God: Michel Houellebecq and Materialist Horror.* Pennsylvania: Pennsylvania State University Press.
Black, R. (2018). *Denied. The Rise and Fall of Climate Contrarianism.* Therealpress.co.uk.
Caudwell, C. (2018). *Culture as Politics. Selected Writings of Christopher Caudwell.* Margolies, D. (ed). London: Pluto Press.
Cook, D. (2018). *Adorno, Foucault and the Critique of the West*. London: Verso.
de Zengotlta, T. (2003). 'Common ground: finding our way back to the enlightenment' *Harper's Magazine*. Retrieved from https://www.thefreelibrary.com/Common+ground%3a+finding+our+way+back+to+the+enlightenment.+.-a099601601
Ehrlich, P. and Ehrlich, A. (2013). 'Can a collapse of global civilization be avoided?' *Proceedings Biological Sciences,* Vol. 8, 280.
Ellul, J. (1965). *The Technological Society.* London: Random House.
Ellul, J. (1973). *Propaganda: The Formation of Men's Attitudes.* Vermont: Random House.
Ereaut, G. and Segnit, N. (2007). 'Warm words II: How the climate story is evolving and the lessons we can learn for encouraging public action'. *Institute for Public Policy Research.* Retrieved from https://www.ippr.org/publications/warm-words-ii-how-the-climate-story-is-evolving-and-the-lessons-we-can-learn-for-encouraging-public-action
Fischer, F. (2017). *Climate Crisis and the Democratic Prospect: Participatory Governance in Sustainable Communities*. Oxford: Oxford University Press.
Fisher, M. (2009). *Capitalist Realism. Is There No Alternative?* Winchester: Zero Books.
Fukayama, F. (1992). *The End of History and the Last Man.* New York: Freedom Press.
Grief, M. (2015). *The Age of the Crisis of Man.* Princeton, NJ: Princeton University Press.
Groys, B. (2014). *The Communist Postscript*. London: Verso.
Gorz, A. (1999). *Reclaiming Work: Beyond the Wage Based Society.* Cambridge: Polity Press.
Hedges, C. (2010). *The Death of the Liberal Class*. New York: Nation Books.
IPCC. (2001) *Climate Change 2001: The Scientific Basis. Contribution of Working Group I to the Third Assessment Report of the Intergovernmental Panel on Climate Change.*

Houghton, J.T., Y. Ding, D.J. Griggs, M. et al. (eds.). Cambridge, UK and New York, NY, USA: Cambridge University Press, 881pp.

Jaeger, C. and Jaeger, J. (2010). *Three Views of Two Degrees*. ECF Working paper, 2/2010.

Jeffries, S. (2021). *Everything, Everywhere, All the Time*. London: Verso.

Lind, M. (2020). *The New Class War: Saving Democracy from the Metropolitan Elite*. London: Atlantic Books.

Lodziak, C. (2000). 'On Explaining Consumption', in Lange, B and Strange, G. (Eds.). *Capital and Class. Environmental Politics: Analyses and Alternatives*. Nottingham: Russell Press, pp. 111–133.

Lukes, S. (2017). *Liberals and Cannibals: The Implications of Diversity*. London: Verso.

Machin, A. (2013). *Negotiating Climate Change: Radical Democracy and the Illusion of Consensus*. London: Zed Books.

Mann, G. and Wainwright, J. (2017). *Climate Leviathan: A Political Theory of Our Planetary Future*. London: Verso.

McGuire, B. (2022). 'Why we should forget about the 1.5C global heating target'. *The Guardian*. Retrieved from https://www.theguardian.com/commentisfree/2022/sep/12/global-heating-fighting-degree-target-2030

Mirowski, P. (2014). *Never Let a Good Crisis go to Waste*. London: Verso.

Nadler, S. (2011). *A Book Forged in Hell. Spinoza's Scandalous Treatise and the Birth of the Secular Age*. Princeton, NJ: Princeton University Press.

Noble, D. (2012). *Debating the End of History*. London: University of Minnesota Press.

North, J. (2017). *Literary Criticism. A Concise Political History*. Harvard: Harvard University Press.

Peckham, M. (1995). *Romanticism and Ideology*. London: Wesleyan University Press.

Pistor, K. (2019). *The Code of Capital. How the Law Creates Wealth and Inequality*. Princeton, NJ: Princeton University Press.

Richardson, K., Steffen, W., Schellenhuber, J., Alcamo, J., Barker, T., Kammen, D., Leemans, R., Liverman, D. Munasinghe, M., Osman-Elasha, B., Stern, N. and Waever, O. (2009). *Climate Change: Global Risks, Challenges and Decision*. University of Copenhagen Synthesis Report. Copenhagen.

Rodney, W. (2018). *Walter Rodney's Russian Revolution: A View from the Third World*. London: Verso.

Roelfs, J. (2003). *Foundations and Public Policy. The Mask of Pluralism*. New York: State University of New York University Press.

Scott, J. (2017). *Against The Grain*. Cambridge: Yale University Press.

Schmidt, J. (2001). *Disciplined Minds. A Critical Look at Salaried Professionals and the Soul-Battering System that Shapes Their Lives*. Maryland: Rowman and Littlefield.

Shaw, C. and Nerlich, B. (2015). 'Metaphor as a mechanism of global climate governance: A study of international policies 1992–2012.' *Ecological Economics*, Vol. 109, 31–40.

Spark, C. (2001). *Hunting Captain Ahab: Psychological Warfare and the Melville Revival*. Ohio: Kent State University Press.

Spufford, F. (2010). *Red Plenty. Inside the Fifties Soviet Dream*. London: Faber and Faber.

Stehr, N. and Grundmann, R. (2012). 'How does knowledge relate to political action?' *Innovation: The European Journal of Social Science Research*, Vol. 25, No. 1, 29–44.

Swaim, B. (2021). 'Hope for the Lost Souls of Liberalism'. *Wall Street Journal*, 2021. Retrieved from https://www.wsj.com/articles/liberalism-crises-political-philosophy-religion-storey-kass-bloom-cancel-culture-war-11631889543

Tocqueville, A. (1835(2003)). *Democracy in America*. London: Penguin Classics.

Vercelli, A. (2017). *Crisis and Sustainability: The Delusion of Free Markets*. London: Palgrave MacMillan.

Wall, D. (2014). *The Sustainable Economics of Elinor Ostrom. Commons, Contestation and Craft*. London: Routledge.

WBGU. (1995). 'Scenario for the derivation of global CO2 reduction targets and implementation strategies. Statement on the occasion of the First Conference of the Parties to the Framework Convention on Climate Change in Berlin'. Retrieved from http://www.wbgu.de/wbgu_sn1995_engl.pdf

WBGU. (1997). 'Targets for climate protection, 1997. A study for the third conference of the parties to the framework convention on climate change in Kyoto'. Retrieved from http://www.wbgu.de/wbgu_sn1997_engl.pdf

WBGU. (2003). 'Climate protection strategies for the 21st century. Kyoto and beyond'. Retrieved from http://www.wbgu.de/wbgu_sn2003_engl.pdf

WBGU. (2009). 'Solving the climate dilemma: The budget approach'. Retrieved from https://www.wbgu.de/fileadmin/user_upload/wbgu/publikationen/sondergutachten/sg2009/pdf/wbgu_sn2009_en.pdf

White, C. (2013). *The Science Delusion. Asking the Big Questions in a Culture of Easy Answers*. London: Melville House.

Whyte, H. (1973). *Metahistory: The Historical Imagination in Nineteenth Century Europe*. Baltimore, MD: John Hopkins University Press.

Winner, L. (1986). *The Whale and the Reactor. A Search for Limits in an Age of High Technology*. Chicago: University of Chicago Press.

Yohe, G. and Toth, F.L. (2000). 'Adaptation and the Guardrail approach to tolerable climate change'. *Climatic Change,* Vol. 45, 103–128.

Zizek, S. (2011). *Living In the End Times*. London: Verso.

2 Guardrail 1: Climate change is not a challenge to individualism

I am free to choose whether or not I engage in carbon intensive activities. You are free to choose whether or not you engage in carbon intensive activities. We are free to choose whether or not we engage in carbon intensive activities. Ultimately it is through our individual, rational free choices that climate change is solved. The state and corporations will respond to the signals they receive from individual choices, and enable the change that individuals demand. It is widely (but not uniformly) believed that technologies alone will not be sufficient to reduce greenhouse gas emissions in line with internationally agreed targets. It will also be necessary for individuals in Western liberal democracies to voluntarily choose to change their behaviours or lifestyles. The International Energy Agency (IEA) reported in 2021 that behaviour change is required for emission reductions and that most of this will come through 'consumer adoption of low-carbon technologies' (IEA, 2021). It is expected that individuals will freely choose these behaviours upon the correct provision of the correct information. The UK Committee on Climate Change has explained that the provision of better public information will allow individuals to see the benefits from making low carbon choices (Committee on Climate Change, 2019: 196). The IPCC highlights the importance of individual choice with the slogan 'Every action matters, every bit of warming matters, every year matters, every choice matters' (IPCC, 2019). Ex US Vice President Al Gore implores people to 'Use your voice, use your vote, use your choice' (Baker, 2020). Such exhortations appear to have little impact on the willingness of individuals to exercise their agency on this topic. A 2022 US poll shows a decline since 2019 in the numbers of Americans who believe their personal choices will affect the trajectory of climate change. In 2019, around 2/3rds of respondents believed their actions had an impact on the climate. This had dropped to 50% by 2022 (Fingerhut and Dolby, 2022). The poll confirmed findings from other polls – people want governments and businesses to lead on climate action (Oliver Wyman Forum, 2021). Jonathan Foley, Executive Director of The Drawdown project, and, according to the Drawdown Project website 'a world-renowned environmental scientist, sustainability expert, author, and public speaker' responded to news of this decline in enthusiasm for behaviour change by arguing 'Individuals should feel empowered to make climate-driven decisions' (Huckins, 2022). These sentiments speak to an ideal of public participation at the heart of the climate debate (Middlemiss, 2014). This may be

DOI: 10.4324/9780429463488-3

participation in deliberative fora (e.g. climate assemblies), membership of community energy schemes, or joining a climate campaign. There is in the liberal imagination what Middlemiss describes as 'an idealised participatory subject fit to build a sustainable future, an individual that is willing to work with others to reach decisions or modify their lifestyle for the greater good' (Middlemiss, 2014: 8). It is difficult to build the enthusiasm for and sense of belief in these sorts of communal activities in liberal societies, wherein people's sense of identity is based on their individual experience rather than their relationships with others (ibid). There is an additional challenge in building communal action at the local scale. Communities have high barriers to entry and exit, they are by definition closed entities (Lindberg, 2018). If they are open to all, regardless of belief, values and identity, then they are a very weak community, there is little to differentiate that group and those activities from their heterogenous surroundings. Liberal societies define themselves in part by their openness, and closed communities are a challenge to this aspiration of openness.

2.1 A visit to the circus

The 'climate camp' protest took place in 2007, in fields near to Heathrow airport, on the Western outskirts of London. Police helicopters flew over the camp ceaselessly. The site was encircled by floodlights and police video cameras, filming all activity within the camp. Anyone leaving and entering was photographed by the police. The camp was big news in the UK (GW News, 2007), as it was essentially a bunch of eco-activists taking over land right next to one of the world's primary transport hubs, a critical node in the circulation of capital and people.

Heathrow airport is near large working class areas in Drayton and Hayes (wealthy people wouldn't freely choose to live so close to such noise and pollution). On the second day of the camp, I had been sitting near the entrance and saw a young person enter the camp who looked very out of place – he didn't have the clothes and self-assured manner of the upper middle class anarchists who were running the camp. After a short time, I saw the young man leave and 30 minutes later return with a friend. It was by now obvious that these young men were from the housing estate just a few hundred yards across the road from this large encampment. The new visitor climbed over the gate into the field, took a few steps in to the camp, looked round, then turned to his friend in complete puzzlement and asked, 'What is this, a fucking circus?'

These young men were of a similar age to the protestors, living in a world facing the same crisis, in a country with a shared language and history, right next door to a high-profile event, and had absolutely no idea why it was happening. (And if, for whatever unfathomable reasons, the climate anarchists had through free choice found themselves lost in the midst of the housing estates of Drayton, they would have been as equally bewildered and may well have asked, 'What is this, a Mike Leigh film set?')

This type of society, where people in the same geographical vicinity have no shared experiences and no common understanding of the world is, of course, something to be celebrated in liberal societies. Technology and money have freed us from geography, custom and obligation to others. Whilst at the international scale, 'climate change can only be dealt with by unparalleled levels of global cooperation' (United Nations Security Council, 2021), at the scale of our lived experiences, communal obligations are not necessary for effective climate action.

Though we barely know the people who live around us, we are expected to come together and work together on agreed strategies for mitigating climate change. But, in line with liberalism's insistence on the separation of problems into distinct units to be solved individually, this collective action is encouraged only as a piecemeal effort to co-opt social energy into delivery of a low carbon liberalism.

These sentiments speak to an ideal of public participation at the heart of the climate debate (Middlemiss, 2014). This may be participation in deliberative fora (e.g. climate assemblies), membership of community energy schemes, or joining a climate campaign. There is in the liberal imagination what Middlemiss describes as 'an idealised participatory subject fit to build a sustainable future, an individual that is willing to work with others to reach decisions or modify their lifestyle for the greater good' (Middlemiss, 2014: 8). It is difficult to build the enthusiasm for, and sense of belief in, these sorts of communal activities in liberal societies, wherein people's sense of identity is based on their individual experience rather than their relationships with others (ibid). There is an additional challenge in building communal action at the local scale. Communities have high barriers to entry and exit, they are by definition closed entities (Lindberg, 2018). If they are open to all, regardless of belief, values and identity, then they are a very weak community, there is little to differentiate that group and those activities from their heterogenous surroundings. Liberal societies define themselves in part by their openness, and closed communities are a challenge to this aspiration of openness.

2.2   Creating the climate individual

To what extent can the individual be abstracted from their social and historical circumstances? Perceiving behaviour as a product of individual preference is a basic principle of the liberal worldview (Ikeler et al., 2022). Whilst seeing systemic issues as part of the reason for climate inaction, authors such as Gifford see the individual's failings as an issue separate from systemic failings. These problems with the individual are to be tackled separately from the systemic problems. The problem is that 'too few global citizens engaged in high-greenhouse-gas-emitting behaviour are engaged in enough mitigating behaviour' and the solution is for 'psychologists to work with other scientists, technical experts, and policymakers to help citizens overcome these psychological barriers'. These psychological failings include 'limited cognition about the problem', 'ideological worldviews that tend to preclude pro-environmental attitudes and behaviour' and 'discredence toward experts and authorities' (Gifford, 2011). It is perhaps worth noting that three years later the author co-authored a paper on the same subject, this time noting that understanding pro-environmental concern and behaviour is far more complex than previously thought and researchers should seek to understand how these factors interact (Gifford and Andreas, 2014). Urry, writing in 2010, realised slowing climate change would require a 'reorganisation of social life but there are no decent sociological analyses of how to bring about this change' (Urry, 2010: 89). We are still awaiting those analyses in 2023, though in 2015 Brulle and Dunlap produced an impressive edited collection drawing together a number of sociological analyses, albeit without any efforts to describe an alternative future (Brulle and Dunlap, 2015). In that edited collection Erhardt- Martinez et al. (2015) talk of 'the nested nature of social systems', an understanding of which can help 'shed light on the interplay between actions at different levels of society' leading to the conclusion that mitigation requires a '"both/and" solution that engages actors at all levels to change both individual and household practices as well as implement organisational, municipal, state national and international policies' (2015: 202). In such pluralist views there is no conflict, but simply the force of reason, rational behaviour choices and the belief a sense of a shared endeavour will unite us all.

In Gifford's article, the types of changes people are being asked to take are piecemeal, irrelevant to mitigation at the individual scale, inconvenient, ahistorical and offer little promise of immediate improvement in the person's life. These are the standard behaviours (installing solar panels, using the bus, buying energy efficient appliances). The problem is not the social structures in which these decisions are to be freely adopted, nor even the behaviours themselves, but a fault with the minds of the proletariat. Gifford does list "ideology" and "worldviews" as two separate barriers to action (though not bourgeois liberalism), but then again positions this not as a structural issue but a fault with the individual for having the incorrect worldview and ideology, though the author does not describe what the correct ideology or worldview is. To be fair to the author, a belief in techno-salvation is listed as a barrier to change, expecting technology to solve the issue, though the actions advised by the author are largely dependent on a belief in technological change as progress towards a better tomorrow (e.g. solar panels and improved energy efficiency).

A sociological analysis tends to the view that individuals matter only in so much as they represent the point of intersection for multiple practices (Ehrhardt-Martinez et al., 2015: 104) or as the otherwise inert carriers of discourses that originate from outside the individual. Writing in 1782, as the idea of the individual as someone in charge of her own destiny began to take hold, de Crevecoeur was able to see that 'we are nothing but what we derive from the air we breathe, the climate we inhabit, the government we obey, the system of religion we profess, and the nature of our employment' (de Crevecouer, 1782). This stark environmental determinism is not often to be found in social science research on climate change. Here, one is more likely to find individuals being recognised as actually existing entities with agency that can be 'mobilised to leverage lower emissions' (Ehrhardt-Martinez et al., 2015: 104). The goal of social science research and public communication efforts is to exploit language to motivate these agentic people to choose to live low carbon lives and support low carbon policies. This agenda has been described as an overly individualistic approach to research (Brulle and Dunlap, 2015: 8) which downplays the role of social context in shaping consumption behaviour 'in favour of a programmatic emphasis on individual consumption choices' (Ehrhardt-Martinez et al., 2015: 227). Economics, the most widely represented discipline in social science research, (Brulee and Dunlap, 2015: 8), sees the social world as made up of individuals rationally maximising the satisfaction of their desires. Any attempt to promote climate-friendly behaviour will have to be aligned with the drive to satisfy our strongest desires in the most efficient and convenient way possible. The one-time economist (and now politician) Per Epsne Stoknes, in his book on the psychology of climate inaction writes that

> the human animal comes with the powerful neural wiring of preferring self-interest (me and us), status, imitation, short term and spectacular risk. Taken together they contribute a lot of the answer at the individual level as to why more isn't happening.
>
> (2015: 34)

But these are ultimately social concerns. For example, imitation is (potentially) an important lever for climate action, as revealed in research on second order beliefs. Second order beliefs are beliefs we have about what other people think and believe. People generally underestimate how much other people worry about climate change and their level of support for climate policies (Mildenberger and Tingley, 2017). Providing people with accurate information about other people's levels of support and concern increased their own levels of support for climate policies.

It is increasingly clear that escalating climate impacts will not be a significant factor in any public awakening and mass mobilisation of the people for building a new future. It is not as though places that experience weather extremes associated with climate change subsequently become centres of radical climate activism. Even where the increasingly inescapable experience of the changing climate does increase climate concern, nothing much changes in terms of behaviour. Research has shown that climate change beliefs have only a small to moderate effect on the extent to which people are willing to act in climate-friendly ways (Hornsey et al., 2016). The high levels of concern about the environment have been reported as consistent in general across most nations for several decades. However, this does not translate into willingness to pay to mitigate climate change (Shwom et al., 2015: 274), and research from the UK indicates a majority perceive climate change to be a collective problem that is not effectively addressed by the behavioural changes of individuals (Lorenzoni, Nicholson-Cole and Whitmarsh, 2007).

But behaviour in bourgeois democracies is not a straightforward result of personal preferences. Middlemiss identifies how the competition that characterises education and work promote individualism as all are pitted against all (Middlemiss, 2014). The ongoing work of competing against one and all, suggests Middlemiss, leaves us little time or energy for the kind of collective action needed to bring about change in the world.

There are many challenges to finding the time and energy to realise our climate agency. Increased productivity is the holy grail of capitalist economies, but as a result of virtually unlimited productivity, we are faced with a need to get rid of the product. Hence, adjusting to work, and the consumption of the goods and services created by work, becomes one of the organising principles of daily life in Western societies (Baudrillard, 1970: 38). This perpetual 'mad dance of accumulation' (Harvey,

2019, cited Swyngedouw, 2022: 907) provides little scope for exercising agency that opposes this treadmill. Another limit on agency is wealth – the agency for ecological action is largely limited to wealthier societies. These limits are especially evident in societies dominated by work, and the building of identity and status through work – it is not the actions of its employees that explain the corporation's actions; on the contrary, corporation's actions explain the actions of its employees (Malm, 2018: 102). Money, power, manufactured consent and political calculations all play a role in constraining agency (ibid.). Caudwell, starting with Freud as his exemplar and then expanding out to the whole field of psychology, argued psychologists assume there is a pure human nature that is frustrated, blocked or distorted by social relations (2018: 24). In climate change psychology, there is a drive to separate out psychology from other elements, and analyse the psychological dimension of individual behaviour through experiments and surveys. Having abstracted humanity from the reality of its social nature into individual units, it then breaks down the individual into separate components to understand why the individual is not behaving correctly.

Psychology, according to Brulle and Dunlap, though also primarily concerned with the individual, at least recognises the limits to human's ability to act rationally (2015: 9) and is a discipline which has made valuable contributions to our understanding of the factors that influence reactions to mitigation and adaptation (ibid: 10). Shove has criticised the 'Attitude, Behaviour, Choice' model underpinning much of the psychology research on climate change, which assumes individual behaviour change will be a driver of effective mitigation (Shove, 2010). This approach is, according to Shove, born of the dominance of economics and psychology in the climate social sciences and exhibits 'a strikingly limited understanding of the social world and how it changes' (Shove, 2010: 1273), and has hence proved ineffectual in bringing about change. Instead, 'behaviour should be considered less the result of an individual's beliefs and desires, and more as being rooted in the logic of practices, routines and habit' (Erhardt-Martinez, 2015: 104).

Some writers look to evolutionary psychology to explain individual inaction on climate. From this deterministic perspective, humans have not evolved to deal with long-term and geographically distant problems (Stoknes, 2015; Marshall, 2015). Our early experiences of trying to survive in harsh environments without the benefits of modern technology have led us to become selfish and greedy, concerned only with satisfying our own immediate needs (Harris, 2015). This, tellingly, is a good description of the most lauded of all liberal types – the entrepreneur, the liberated individual on a solo quest for self-actualisation (Lordon, 2014: 3). The environmentally conscious entrepreneur (such as Richard Branson, Bill Gates and no end of other white males) is enacting a kind of "drama" which seeks to perform the possibility of capitalism and a safe climate as being mutually compatible, with the bourgeois entrepreneur as the central protagonist in the realisation of this vision (Prudham, 2009).

For those arguing our destructive habits are hardwired, the explanation for the acceleration in the destruction of the biosphere is simply a matter of numbers. We have always everywhere been the same, there are just more of us now so the damage is that much worse (Mars-Jones, 2016). We will aim to accrue the most calories possible in the least calorific way. This leads to a pattern of megafauna being driven to extinction whenever humans appear. Harari uses the example of Australia to make this point, where the megafauna disappeared 45,000 years ago, just when humans arrived on the continent (Harari, 2011: 73). The most famous and more recent example of this phenomenon is the dodo. More recently still, the passenger pigeon in the US, which flew in flocks billions strong and many miles long was made extinct by overhunting in 1914. Industrialisation plus increased numbers of humans would seem to give the rest of the ecosphere little chance of long-term survival, if we are indeed prisoners of our own destructive tendencies. All life will seek to reproduce itself with the minimum effort possible. Why are the birds in my garden so attracted to the high fat suet balls bird feeder I can see from where I am sat now? Because it is the least calorific way of sourcing calories. To return to Harari's thesis, the reason the history of humanity is in large part the story of the extinction of megafauna is because megafauna is an easy source of large volumes of calories. Why have humans exploited fossil fuels so intensely? Because it is the easiest way to source large volumes of energy. A single litre of petrol used today needed about 25 metric tons of ancient marine life as precursor material (Dukes, 2003, cited Mitchell, 2013: 18). We are still concerned with extracting the most out of life with the least effort, which today takes the form of convenience, and convenience is a major driver of growing energy use (Shove, 2003). Critical thinkers such as Adorno recognise the existence of instinctive drives but do not identify them as a universal foundation of human nature that precedes socialisation. Rather it is socialisation under modernity that reifies these elements, that raises them up into a social good and desirable feature of a person's identity (Adorno, 1942). Hence it is our social being under liberalism that is the barrier to climate action rather than an enduring and universal feature of humanity.

Neither evolutionary psychology nor socialisation leaves much room for speaking of individuality or free will as the locus of climate action. In a wide-ranging review of social science research into climate communication, Moser finds little evidence to support the validity of using the individual as the primary unit of analysis to generate strategies for motivating behaviour change, and instead describes the complex 'intra-psychic, interpersonal, social, cultural and political-economic dynamics that shape people's responses to climate change' (Moser, 2016: 353). Similarly, Chapman et al. find a role for the individual in seeking to understand our emotional responses to climate change, but only as one element amongst many, including 'rudimentary feeling-states combined with a range of cognitive appraisals of context, the self, and others, as well as (multiple) potential motivational impetuses' (Chapman, Lickel and Markowitz, 2017: 851).

The social is real, it is the individual that is the abstraction. Socialisation and historical circumstance go all the way down. There is no human nature that precedes this (Rorty, 1990: 5). The self is simply whatever acculturation makes of it. Agency lies in the interpersonal realm, not with the individual (Burr, 1998). Take away everything that the social contributes to the individual's nature and what are you left with? Contrary to liberalism's claims to treat each individual as an entity possessed of equal rights before the law, as a potentially rational, creative and moral creature capable of self-knowledge, social knowledge and self-management (Spark, 2002: 14), the individual is little more than a bourgeoise illusion which seeks to deny the truth that all reality and progress is created through social relations (Caudwell, 2018: 20).

How does the wishful thinking about the role of individuals working cooperatively with liberal institutions play out in the real world? In 2015, Nancy Young was Vice President for environmental affairs with Airlines for America, the United States' largest airline industry lobby group. I have never met Nancy Young, but as she is Vice President of something I have to assume she is a very capable person. I would go further. Nancy is a successful woman who is working to build a sustainable future. Add in that the industry Nancy is working in is one essential to globalisation, and connecting humans across the world, and you have the archetype of the modern liberal – she symbolises the realisation of some many of the promises made by the liberal idea of progress. Nancy Young – as an "ideal type" – is smashing it with regard to what human progress is all about. When, in a 2015 interview, Nancy Young was asked about the contribution airlines are making to rising emissions of greenhouse gases, she replied 'What do you want us to do? Do you want us to suppress demand for flying? Do you want people to not see grandparents or bring critical goods to earthquake zones?' (Holthaus, 2015). I don't know what percentage of flights are taken so that grandchildren and grandparents can spend time together, or what percentage are used for bringing critical goods to earthquake zones. I suspect it is quite small. What is clear is that the contribution to climate change from flying is being made by a tiny minority of the world's population. It has been calculated that only around 3% of the global population flew in 2017, and at most, only about 18% have ever flown (Gosling, 2020).

The interesting point about Nancy Young's response goes beyond the issue of flying, or Nancy Young, because if Nancy Young hadn't offered that response she wouldn't be in the role, which would instead be occupied by someone willing to give the answer she offered. Wherever you turn, you will find a Nancy Young, pulling down the shutters on reasoned debate, and defensively latching onto a confused justification for why their industry, their little corner of the liberal edifice, must remain exempt from reconsideration. It is in these discourses we see liberalism cutting off our escape route out of climate catastrophe. It feels important to be clear, the examination of individual speech and writings in this book are not attacks on the individuals themselves. In modern societies, people are nothing more than the 'precipitates of language' (Peckham, 1995), little more than the carriers of institutional discourses. There is no self outside of language, for how can we become autobiographical without language (White, 2013: 135). The Norwegian author Knausgaard is moved to write that 'the world flows through the individual to such a remarkable extent that one has to wonder if anything like individuality even exists and if it does in what way it might manifest itself' (Knausgaard, 2018). To "the world" we might also append, "the word".

2.3   The search for individual free will

Sociological analyses suggest that our decisions are much more conditioned by social conditioning and structural issues than individual free will. The claim of rational individual preference is simply false: our response to social norms is to a large extent 'blind, and compulsive, mechanical and even unconscious' (Elster, cited Lukes, 2017: 4). In his review of consciousness and decision-making, Nunn concludes that 'as far as conscious experience goes, there is nothing but society' (Nunn, 2005: 190). Though, as we have seen above, the idea of the individual is useful for analytical and communication purposes, the existence and nature of the individual is a highly contested one, rather than an unarguable prior for analysis of human activity. Social norms and influence have been found to change people's environmental behaviour, but only on minor issues and in minor ways. For example, sharing information with hotel guests about how many other guests are using their towels more than once without needing them washing changes people's willingness to reuse hotel towels (Goldstein, Caldini and Griskevicus, 2008). Such trivial adjustments in behaviour are supportive of those seeking to promote a non-conflictual transition to a less polluting form of bourgeois liberalism, but have nothing to offer the fight of our lives we now find ourselves in. As the African proverb explains, when the music changes, so does the dance. There is no essential human nature. Under normal conditions everyone is more or less good, or, at least, ordinary. But tempt them and they may suddenly change. Much of the work on testing the impact of messaging on people, in assuming the words chosen can have some kind of impact on people's attitudes (if not their behaviour), must then also accept that the other discourses of modernity have an equal impact. Except the discourses promoting energy intensive behaviour are much more pervasive and persuasive than those promoting low energy lifestyles.

Evolutionary psychology has not created monsters of us, desperately fighting each other to maximise our own satisfaction. It is liberal modernity which provides the conditions which reward those behaviours. Liberalism has since the 20th century accelerated a move from communal to associative forms of action. Associative action begins with freedom from the bonds of community and tradition but, according to Dickens, leads to our imprisonment in dehumanised systems, where freedom has decayed into a series of private value choices and goal-oriented actions (Dickens, 1992: 28). The individual is thoroughly embedded in and conditioned by exchange relations under late capitalism and it is these present conditions, not prehistory, that is responsible for encouraging and praising selfishness. We are just the effects of forces of which we are largely unaware and over which, for that very reason, we exercise little or no control (Cook, 2018: 61). But those forces are not evolutionary psychology, but the forces of social control. Hence, such individualism as exists is only present in our role as actors within the terms set by power, by the act of inscribing on ourselves the limits, constraints and perceptions we are socialised into seeing and accepting as the truth by doctors, teachers, journalists and other experts. Power relations have us and are exercised through us (Weberman, cited Cook, 2015: 70). The individual, as much as it exists, is weak, is produced by power to be weak (ibid.: 73). However, liberal capitalism requires a degree of autonomy from the bourgeois so they can reproduce capital without expensive supervision, but this is largely reduced to individuals defining their interest and then defending that interest against others (Adorno cited Cook, 2015: 77). If we actually understood how much we are created by social norms that predate our existence, we would despair as the illusion we are individuals exercising free will would be destroyed and we would see that we are, individually, nothing.

There is a confluence of thinking around the individual that begins to overlap with much Buddhist thinking, which has for 3000 years flourished as a philosophy that rejects the idea of identity and the separateness and reductionist worldview which defines bourgeois liberalism. Adorno's discussion of negative dialectics looks to nonidentity, thinking which eschews the idea of concepts, identity and any sort of fixed reality (Cook, 2015: 22). Of course, liberalism has no concern for us as individuals. The individual is a pragmatic unit of analysis for researchers, it is helpful for power to have people atomised and at each other's throats, and it is useful for capital to have 7 billion people seeking identity and connection through consumption. But there is no intrinsic concern for the suffering that will be experienced from environmental degradation in the pursuit of these ends.

2.4   Hegemonic climate communication

Brulle et al. argue that a large proportion of the population have fixed views that are unlikely to change. There is another section of the population – highly educated – who are attentive to and heavily swayed by elite cues (Brulle, Carmichael and Jenkins, 2012). So, rather than the individual's limited cognition or lack of knowledge about the science of climate change being the focus of analysis, the issue to be fixed is actually classes of people, their position in this historical and social moment.

Despite the professed concern for the individual, dominant liberal climate discourses are hegemonic, and in fact, are focused on reaching a particular (influential) class of the population, rather than the whole population, in order for the governing class to win the consent of the governed through peaceful rather than violent means. The relatively small numbers of rich and wealthy actors supplement their reach by financing an intellectual class, through foundations and think-tanks – to do their hegemonic work for them (Roelofs, 2003: 3). The recognition of a van-guard or hegemonic bloc that will lead on defining the trajectory of climate action is however absent from institutional discourse on the role of the individual. Despite talk of 'the individual', institutional discourse treats the individual in a remark-ably homogeneous way. Perry Anderson's history of the idea and use of the term 'hegemony' (Anderson, 2017) records how Isocrates, in the 4th century BC, used the term to describe 'the power of the word over all things' (2017: 5). The word 'hegemon' was not to reappear until the mid-19th century and the (re)emergence of democracy in Europe (ibid.: 6). In this reincarnation, it was originally used to describe relationships between states, but at the turn of the 20th century, it began to be used to describe political relationships within states (ibid.: 13). The next evolu-tion saw the term as used by Lenin to describe control by the working class and their proletarian allies over Russian society after the revolution, Gramsci generalised the term to describe stable forms of rule by any social class (ibid.: 19). Gramsci also expanded the forms of rule under hegemony to include the use of violence, which has the appearance of being backed by the consent of the majority, as expressed 'by the organs of opinion' (Gramsci, cited Anderson, 2017: 20). What Gramsci wanted to explain was the moral consent of the dominated to their own domination, or in our instance, the moral consent of the damned to their own damnation. Such consent is only possible where the subjugated internalise the values and descriptions of the world of the groups doing the subjecting. This was only possible in societies with

> a well-equipped and long-established intellectual strata in developing and diffusing the ideas of the ruling order downwards through subordinate classes. The other element was a dense network of voluntary associations in one way or another purveying the outlook of capital
>
> (Anderson, 2017: 21–22)

Together, these actors can construct a new common sense, that identifies freedom with the market and order with moral tradition, suitable for popular consumption (Anderson, 2017: 87). This is the political terrain targeted by organisations such as the Central Intelligence Agency, the non-communist left. These moderate reform movements, by setting the boundaries of acceptable left-wing thought, are essential to safeguarding liberal interests (Roelofs, 2003: 11).

2.5   Conclusion

The individual is both the villain and the heroine of the liberal climate story. Individuals possess the power to change the world, but are hampered in exercising this power by evolutionary psychology (which has made us greedy, selfish, tribal and short-sighted), and a failure to heed or hear the messages and prompts which will ensure individuals freely choose to make the correct choices. The social does not really exist except as something that emerges at odd moments, such as when individuals choose to engage in participatory decision-making or join an energy collective. Psychological research into the climate subject often requires that the subject be removed from the social context so that the effect of message testing and other psychological tests can be better measured and analysed. This individualisation of the climate subject is in direct contrast to the body of work which insists the individual is an abstraction. From this perspective, our real nature is social, and historical and social conditions are the real drivers of, and limitations on, our agency.

References

Adorno, T. (1942). Theses on Need. Retrieved from http://freudians.org/wp-content/uploads/2015/08/theodor-w-adorno-theses-on-need-1.pdf
Anderson, P. (2017). *The H-Word. The Peripeteia of Hegemony.* London: Verso.
Baker, K.C. (2020). 'Al Gore Says "We Have the Solutions We Need to Solve the Climate Crisis" on Earth Day 2020'. *People.* Retrieved from https://people.com/human-interest/al-gore-says-we-have-the-solutions-we-need-to-solve-the-climate-crisis-on-earth-day-2020/
Brulle, R., Carmichael, J. and Jenkins, C. (2012). 'Shifting public opinion on climate change: an empirical assessment of factors influencing concern over climate change in the U.S., 2002–2010'. *Climatic Change,* Vol. 114, 169–188.

Brulle, R. and Dunlap, R. (2015). Sociology and global climate change', in Brulle, R. and Dunlap, R. (eds). *Climate Change and Society. Sociological Perspectives.* Oxford: Oxford University Press, pp. 1–31.

Burr, V. (1998). 'Overview: Realism, relativism, social constructionism, discourse', in Parker, I. (ed). *Social Constructionism, Discourse and Realism.* London: Sage, pp. 11–13.

Caudwell, C. (2018). *Culture as Politics.* Margolies, D. (ed). London: Polity Press.

Chapman, D.A., Lickel, B. and Markowitz, E.M. (2017). 'Reassessing emotion in climate change communication'. *Nature Climate Change*, Vol. 7, 850–852.

Committee on Climate Change. (2019). Net Zero. The UK's contribution to stopping global warming. Retrieved from https://www.theccc.org.uk/publication/net-zero-the-uks-contribution-to-stopping-global-warming/

Cook, D. (2018). *Adorno, Foucault and the Critique of the West.* London: Verso.

Erhardt-Martinez, K., Schor, J., Abrahamse, W. et al. (2015). 'Consumption and climate change', in Dunlap, R. and Brulle. R. (eds). *Climate Change and Society: Sociological Perspectives.* Oxford: Oxford University Press, pp. 93–126.

Erhardt- Martinez, K; Rudel, T.K, Norgaard, K.M and Broadbent, J. (2015). 'Mitigating climate change', in Dunlap, R. and Brulle. R. (eds), *Climate Change and Society: Sociological Perspectives.* Oxford: Oxford University Press, pp. 199–234.

European Commission (2022). Energy communities bring renewable power to the people. Retrieved from https://ec.europa.eu/research-and-innovation/en/horizon-magazine/energy-communities-bring-renewable-power-people

Fingerhut H. and Dolby, N. (2022). AP-NORC poll: Many in US doubt their own impact on climate. *Associated Press News.* Retrieved from https://apnews.com/article/inflation-science-technology-climate-and-environment-9038738cbda161f43e5d0d2fbd5e738b

Gifford, R. (2011). 'The dragons of inaction: Psychological barriers that limit climate change mitigation and adaptation'. *American Psychologist*, Vol. 66, No. 4, 290–302.

Gifford, R. and Andreas, N. (2014). 'Personal and social factors that influence pro-environmental concern and behaviour: A review.' *International Journal of Psychology*, Vol. 49, No. 3, 141–157.

Goldstein, N., Cialdini, R. and Griskevicius, V. (2008). 'A room with a viewpoint: Using social norms to motivate environmental conservation in hotels.' *Journal of Consumer Research*, Vol. 35, 472–482.

Gössling, S. and Humpe, A. (2020). 'The global scale, distribution and growth of aviation: Implications for climate change'. *Global Environmental Change*, Vol. 65.

GW News (2007). Summary of 5 weeks of Heathrow protest coverage from GW News. Retrieved from http://geosci.uchicago.edu/~rob/heathrow.html

Harari, Y.N. (2011). *Sapiens. A Brief History of Humankind.* London: Vintage.

Holthaus, E. (2015). "Just Plane Wrong". *Slate.* Retrieved from http://www.slate.com/articles/business/moneybox/2015/06/air_travel_and_climate_regulation_why_the_epa_might_let_big_aviation_off.html

Huckins, G. (2022). The Psychology of Inspiring Everyday Climate Action. Wired. Retrieved from https://www.wired.com/story/the-psychology-of-inspiring-everyday-climate-action/

Hornsey, M., Harris, E., Bain, P., Kelly S. and Fielding, K. (2016). 'Meta-analyses of the determinants and outcomes of belief in climate change'. *Nature Climate Change*, Vol. 6, pp. 622–626.

Ikeler, P., Sernatinger, A. and Cass, E. (2022). 'When common sense fails'. *Spectre Journal.* Retrieved from https://spectrejournal.com/when-common-sense-fails/

International Energy Agency (2021). Net Zero by 2050. A Roadmap for the Global Energy Sector. Retrieved from https://iea.blob.core.windows.net/assets/4719e321-

6d3d-41a2-bd6b-461ad2f850a8/NetZeroby2050-ARoadmapfortheGlobalEnergySector. pdf

IPCC (2019). AR6 Assessment Cycle Fold-Out Card. Retrieved from https://www.ipcc.ch/ site/assets/uploads/2020/11/FOLDOUT_CARD2019.pdf

Irene Lorenzoni, I., Sophie Nicholson-Cole, S. and Lorraine Whitmarsh, L. (2007). 'Barriers perceived to engaging with climate change among the UK public and their policy implications'. *Global Environmental Change*, Vol. 17, No. 3–4, 445–459.

Knausgaard, K. (2018). *The End*. London: Penguin

Lindberg, E. (2018). Look and See; Listen and Hear: Wendell Berry and the Contradictions of our Climate. *Resilience*, Retrieved from https://www.resilience.org/stories/2018-07-23/ look-and-see-listen-and-hear-wendell-berry-and-the-contradictions-of-our-climate/

Lordon, F (2014). *Willing Slaves of Capital. Spinoza and Marx on Desire*. London: Verso.

Lukes, S. (2017) *Liberals and Cannibals: The Implications of Diversity*. London: Verso.

Malm, A. (2018). *The Progress of This Storm*. London: Verso.

Mars-Jones, A. (2016). Chop and Burn. *London Review of Books*, Vol. 38, No. 15, 23–25. Retrieved from https://www.lrb.co.uk/the-paper/v38/n15

Middlemiss, L. (2014): 'Individualised or participatory? Exploring late-modern identity and sustainable development', *Environmental Politics*, Vol. 23, No. 6, 929–946.

Mildenberger, M., and Tingley, D. (2019). Beliefs about climate beliefs: The importance of second-order opinions for climate politics. *British Journal of Political Science*, Vol. 49, No. 4, 1279–1307.

Mitchell, T. (2013). *Carbon Democracy. Political Power in the Age of Oil*. London: Verso.

Moser, S (2016). What more is there to say? Reflections on climate change communication research and practice in the second decade of the 21st century. *WIREs – Climate Change*, Vol. 7, No. 3, 345–369.

Nunn, C. (2005). *De La Mettrie's Ghost: The Story of Decisions*. London: Palgrave Macmillan.

Oliver Wyman Forum. (2021). Consumers Say National Governments Should Lead on Climate Change. Retrieved from https://www.oliverwymanforum.com/climate-sustainability/2021/sep/consumers-say-national-governments-should-lead-on-climate-change.html

Peckham, M. (1995). *Romanticism and Ideology*. London: Wesleyan University Press.

Rorty, R. (1990). *Objectivity, Relativism and Truth*. Cambridge: Cambridge University Press.

Shove E. (2010). Beyond the ABC: climate change policy and theories of social change. *Environment and Planning A: Economy and Space*, Vol. 42, No. 6, 1273–1285.

Shove, E. (2003). 'Converging conventions of comfort, cleanliness and convenience.' *Journal of Consumer Policy*, Vol. 26, No. 4, 395–418.

Shwom, R., McCright, A., Brechin, R. et al. (2015). 'Public opinion on climate change', in Dunlap, R. and Brulle, R. (eds). *Climate Change and Society*. Oxford: Oxford University Press, pp. 269–299.

Spark, C. (2001). *Hunting Captain Ahab: Psychological Warfare and the Melville Revival*. Kent, OH: Kent State University Press.

Stoknes, P. (2015). *What We Think about When We Try Not to Think about Global Warming: Toward a New Psychology of Climate Action*. Vermont: Chelsea Green Publishing.

3   Guardrail 2: The liberal construction of climate change is universally true

> Western representations of climate risk and acceptable levels of harm sit above ideology, national interests, culture and geography. There is one dangerous limit to climate change. It is the same whoever and wherever you are. Using this target as a policy objective helps coordinate global responses under the one right way of being in the world.

Fanon spoke of the modern collapse of reason and history into all things European (Gordon, 2015: 20). What Fanon is describing here is the deeply held belief that progress, the march into light, the better tomorrow will be, above all else, congruent with the idea of Europe. Any change that is not rooted in the European ontology of science, control and nature as 'other' must be fought, defeated and removed from memory. What this meant for Fanon is nothing less than 'the European ambition to become ontological, that Europe sought to become, "absolute being", a theodicy of Western civilisation and thought as systems that were complete and intrinsically legitimate in all aspects of human life' (ibid.: 20).

Climate change has become yet another means by which Europe and the West has been able to extend its control over distant and subordinate places (Stipple and Paterson, 2007: 162). This liberal global climate regime has become depoliticised but not de-ideologized. Rather, it is ideological through and through (ibid.: 150). As part of the research for my doctoral thesis (Shaw, 2011), I interviewed a researcher with an interest in climate justice. The interviewee described meeting young representatives of the 350.org campaign group at a conference, and recounted the following interaction:

> They sort of were friendly and laid out their thing and said "you know, we have had all these nationwide demonstrations and our starting point was 2 degrees, our starting point is we have to cut emissions, we have to have targets. Emissions have to be stopped at this level, this is our first step. Once we can get a movement aligned around" (2 degrees, or a movement aligned behind whatever target it was they were talking about their idea was), "this is the major thing we need to do, the first step we need to take and after that we can work out how we are going to do it". And I said "frankly in all of these discussions over many years (with indigenous people from the South) the

DOI: 10.4324/9780429463488-4

issue of emissions targets and the issue of 2 degrees and the issue of 350 parts per million has never once come up. In reality when you are talking climate politics with people in the South, this is not where the core of the issue is, I mean, it is really not the core of the issue and it has never come up in our real practical work with our groups in the South". And the two students, at first they thought we were either kidding or we weren't really feeling the climate problem or we weren't of use to them and their movement. Their communication, in a way their sympathy for us, just melted away and they didn't understand what we were trying to waste our time about if we weren't trying to organise the world around 2 degrees or 350 parts per million or whatever it was.

This evangelism for Western world views is embodied in the figure Alden Pyle, in Graham Greene's *The Quiet American*. Pyle is in Indo-China on behalf of the Central Intelligence Agency, enabling terrorist attacks against the communist regime. Though these bombings cause immense suffering, Pyle is happy in his work. Pyle knows with a zealous fervour that he is working for the greater good, a liberal future for everyone. The analogy between the two forms of imperialism is a legitimate one, given that the horror visited upon the people of South East Asia by the West between the 1950s and 1980s is now being replicated globally: 'In its current form, globalised neoliberal capitalist culture now unequivocally threaten humans and nonhumans alike with a ghastly future that demands an unprecedented transformation of modern society's anthropocentric and colonial worldview' (Piccolo, Taylor and Washington et al., 2022).

3.1   Institutional norms and the liberal imperialism of climate change

It is the claim of this chapter that the climate discourse regime set by institutions is characterised by an internationalisation of a response to climate change that is birthed in the West's liberal dominance of the world. The imposition of a liberal construction of climate risk across the globe is made possible through international liberal institutions, and so institutions form the main unit of analysis in this chapter. The concept of institutions is very broad, ranging from formal deliberating bodies engaged in treaty-making, to the informal liaisons among a range of different decision-making and non-decision-making communities and actors (O'Riordan & Jordan, 1999). What defines all these interactions as institutional is the 'presence of some sort of order and guiding principles of social solidarity' – rules and norms – focused around a 'locus of regularised or crystallised principle of conduct that governs a crucial area of social life and that endures over time' (O'Riordan & Jordan, 1999: 346).

Debates about human nature that took place during the 20th century shaped the norms adopted by global institutions such as the UN (Greif, 2015). These institutions were active participants in the discussions promoting the sense of a universal humanity. An overriding norm of international institutions is the search for international agreements on commonly agreed goals. This means climate institutions are defined by norms which assume climate change can only be solved by international agreement on common goals. Liberal economic and governmental reform since the 1970s across a broad sweep of political activity has had the primary goal of transferring decision-making authority to remote technocrats who operate outside of the constraints of formal democratic politics. The capture of science by institutional liberal norms based on a universal view of the liberal human type is a common feature of institutions that evolved to deal with the trans-national environmental problems that have emerged since the latter part of the 20th century.

Bernstein's *The Compromise of Liberal Environmentalism* (2002), provides a detailed genealogy of the genesis of the formation of a liberal environmental hegemony, drawing on primary sources at a time when experience of this institutionalisation was fresh in the mind of key architects. It is a key source for this part of the chapter. Bernstein wants to know,

> At the time when different economic models for managing environmental problems were being debated, including the idea that all natural resources will have to be state controlled and declared the common property of the community, how did liberal environmentalism come to dominate?
>
> (2002: 182)

The downplaying of the social and political changes needed to mitigate climate change in favour of purely technical and scientific responses began in the early 1970s. Bernstein (2002: 29) argues the origins of the compromise of liberal environmentalism can be found at the 1972 Stockholm United Nations conference on the human environment conference. Policy-makers needed a theme to unify the strands developed at Stockholm into a unified programme at the 'Earth Summit' held in Rio de Janeiro in 1992. From here, new thinking developed that attempted to link environment and development into a single framework, hence "sustainable development". This represented a shift in environmental policy to deregulation, cost benefit analysis, and a heavy reliance on market incentives. Bernstein stresses that this decision came from an ideological shift rather than an assessment of policy effectiveness (Bernstein, 2002: 53). In the 1980s, this shift was accelerated in a programme that came to be called neoliberalism, and an acceptance of the established economic order provided the backdrop for the concrete activities of the international environmental institutions and diplomats (Broadhead, 2002: 101). At a 1990 IPCC meeting in Washington, President Bush spoke of climate change as an issue needing solutions '...consistent with growth and free market principles' (Pettenger, 2007: 30). The international liberal institutional complex could only take environmentalism seriously when it was considered in the context of an economic programme that not only encouraged growth but actually demanded it. It was at this point, in the early 1990s, that the compromise of liberal environmentalism became embedded in institutional norms, and liberalism became understood as the solution, rather than the problem (Mair, 2014). Accordingly, the much publicised Labour government's 2007 Stern Report made a conscious decision to frame climate as an economic issue (Willis, 2017).

Bernstein shows that there was a norm-complex of embedded liberalism in the post-World War II era. What happened with the institutionalisation of climate change was exactly what would be expected, given our hypothesis that those setting the parameters of the discourse would have had to draw upon what was already familiar to them to make sense of this new phenomenon. That meant accepting the liberalisation of trade and finance as consistent with, even necessary for, international environmental protection (Bernstein, 2002: 7). This leads to the promotion of market mechanisms such as tradable pollution permits over command-and-control methods such as bans, quotas and so on. Consequently, with the Rio declaration in 1992, the liberal economic order acquired for the first time a normative character in an international instrument relating to the environment (Bernstein, 2002: 118). We will examine how these considerations informed the creation of the IPCC below. For now, it is worth noting that for many of the key actors at the head of these institutions, the belief in liberal market principles is stronger than the belief in the need for climate action. Bernstein writes that the heads of the International Monetary Fund, the World Bank, World Trade Organization and the Organization for Economic Co-Operation and Development all know each other and often move from one organisation to the other (2002, 117). This rather unsurprising state of affairs would perhaps not be worth comment, except that in 2022 the World Bank president David Malpass was asked if he believed in the scientific consensus that the burning of fossil fuels is rapidly and dangerously warming the planet. 'I'm not a scientist,' was Malpass' reply (Thwaites, 2022). This would indicate that a certain scepticism towards claims of climate change as an immediate and urgent crisis is the norm amongst the heads of these economic organisations.

3.2   The communication of liberal institutional norms in climate discourses

Since institutions are constituted upon shared meaning, they are reformed or maintained through discourse (Avigur-Eshel, 2019). They also maintain the social and historical conditions for their own relevance and agency in the world through discourse. Schmidt (2008) and Schmidt and Radaelli (2004) identify two forms of discourse in an institutional context: coordinative and communicative. While coordinative discourse is used among policymakers, communicative discourse 'consists of the individuals and groups involved in the presentation, deliberation, and legitimation of political ideas to the general public' (Schmidt 2008: 310). These communicative discourses prioritise the idea of the individual and the idiographic in their construction of the world. However, the need to centralise discussions of risk and risk responses within the norms of Western liberal institutions means imposing a single risk metric on the whole world, in a sense denying this very individualism and idiographic experience of the world. Hence, climate change forces the critic to confront an inherent contradiction at the heart of liberal claims of a cosmopolitan individualism 'which treats every individual as equal, morally and legally, for when we are discussing this idea of the universal rule for all people, we are, in fact, talking about Europeans' (Lukes, 2017: 6).

3.3   Climate targets and the communication of liberal norms

As noted previously, the idea of a single, universal and knowable dangerous limit is the big bang of climate change, the ultimate creation myth for our time. This claim defines just about everything it is possible to think and say about climate change. If you doubt this, try talking about any plan for doing something about climate change that doesn't have at its core the claim of a knowable single dangerous limit. It is the "once upon a time" of all our climate change stories. And whilst it may, like all good fairy stories, offer some insights into how humans make sense of the world, it is not a product of science, it is not a property of the atmosphere, it is not, in any objective reckoning, true and it is not just. The dangerous limit concept is a cultural artefact which has been constructed under a historically specific set of economic and social conditions by a handful of powerful actors from the global North. That is to say, under different social and economic conditions, we might have expected a different construction of climate change to emerge.

We understand climate change as a phenomenon with a single dangerous limit because of a strategy Mair identifies as 'containment' (2009). Containment is one of two institutional strategies for problem management, the other being what Mair titles 'crisis management'. Containment (the strategy most closely aligned with the guardrails framework employed in this book) includes two practices: boundary setting and tailored framing. Boundary setting concerns the delimitation of the possible. Discourses set the boundaries by acknowledging the possibility of applying alternatives, but only to de-legitimise them. It is in essence a "no-alternative" tactic (Mair, 2009). Containment reinforces the existing ideological framework as a basis for producing public policy. It conveys the message that the challenge is the product of *problems* in implementing (liberal) ideas rather than the ideas themselves (Avigur-Eshel 2014). Working within the containment strategy results in institutions such as the IPCC becoming a conservative organisation predestined for understatement (Powell 2011: 47, cited Dunlap and McCright 307). In reality, the IPCC, in its initial version as a purely scientific organisation (the Advisory Group on Greenhouse Gases [AGGG]) did pose very challenging problems for policymakers. The AGGG was made up almost entirely of scientists. In 1988 it was replaced by the IPCC, which was characterised by a process that brought in governments from across the world, giving them the chance to define the questions to be assessed by the climate scientists and the chance to approve the final report. This meant the IPCC was a much more political institution than the AGGG. Boehmer-Christiansen (1994) has argued that the AGGG was disbanded for being too partisan and policy prescriptive. In other words, authorities were unhappy that decisions about limits to warming were being made without their oversight.

Bernstein records that James Bruce was Secretary of the World Meteorological Organisation (WMO) executive when it decided to set up the IPCC. Bruce said that after a couple of years of the work of AGGG, an unease crept into some governments that climate change was going to have enormous economic repercussions and those governments, in particular the United States, didn't like the idea of these freewheeling scientists pronouncing on the subject. They preferred something with more governmental involvement (Bernstein, 2002: 166). The IPCC was created to impose (liberal) discipline on the scientific debate (Liverman, 2009). It was within the context of constructing a story of climate change that did not challenge the politics of Western liberal imperialism that the idea of a single knowable (and distant) dangerous limit for climate change emerged.

At the United Nations global 'Climate Ambition Summit' in December 2020, it was claimed that 'the science is as urgent as ever and tells us that we need to limit the rise in global temperature to 1.5°C degrees'. Few climate scientists would back the claim that science can define what counts as an acceptable level of climate risk for all 7 billion people on the planet, wherever they live, whatever their circumstances, regardless of their vulnerabilities. For science to define 1.5°C as the upper limit for warming would require science to also be able to explain why the harms currently being experienced at 1°C of warming (IPCC, 2018: 11) are acceptable, but those at 1.5°C are not. Such target setting is a carefully considered element of liberal hegemonic practice and performance. Targets turn the problem into a viable object of decision-making for liberal institutions (Lahde, 2006: 87). Language is being used to justify an ideological choice and make the exercise of power appear to be an objective fact, by substituting discussion of what counts as an acceptable level of climate risk with the incontestable claim of a scientifically derived dangerous limit (Machin, 2013).

The dangerous limit concept is a trade-off between values, ideologies, power, risk and ideas of what is a good life and who gets to live that good life. Liberal ideas of the good life did not come to dominate through being self-evidently better than existing ways of being. Enlightenment liberalism is a fighting creed, armed with a universal account of human nature, and of how societies arise and function, universal notions of human interest and human freedom, and universal prescriptions for education and moral progress, all of which could be established from a scientifically objective and universally attainable point of view (Hollis, 1999: 36, cited Lukes, 2017: 34). Economic liberalism is a deliberate effort to extend the struggle and competition for resources to all domains of existence, all corners of the world (Betty, 2017: 80). Liberalism has made a fetish of the desirability and possibility of a return to a borderless world for capital, making states adopt identical laws regarding property rights, freeing investors from vexatious legislation. 'All that is to be left of the state is patriotic flag-waving nostalgia' (Streeck, 2020: 114).

3.4   The denial of uncertainty and the denial of climate justice

The United Nations Framework on Climate Change Convention (UNFCCC) is an institution governing a multilateral legal instrument specifying a system of norms and rules for avoiding dangerous climate change (Porter & Brown, 1991: 20). Given that there are a number of uncertainties about future climate impacts and potential climate tipping points, these norms and rules come to play a determining role in defining what is true (Johnson & Covello, 1987: 357). Therefore, the 'global reproduction of knowledge is more dependent on the institutions involved than the facts themselves' (Wynne & Jasanoff, 1998: 20; Eden, 2004: 59). The primary role of these institutions is to absorb uncertainty and turn it into fact (Eden, 2004: 59). This consensual validation plays an important role in reaching decisions about what counts as a problem, how problems are represented, and, consequently, the best strategies to be used to solve those problems (Eden, 2004: 50). This problem definition is part of the process for creating meta narratives. Meta narratives are policy narratives in a controversy that embrace the major oppositions to a controversy without slighting any of that opposition (Roe, 1994: 52, cited Hamilton, 2009: 228). These meta narratives birth policy narratives that underwrite and stabilise assumptions for decision-making (Hamilton, 2009: 233).

This is the point at which liberal ideology comes in, to replace the unknowability of our climate future with a story aligned with the interests of the ruling class. Ideology lies in the paradoxes and contradictions of the political system at the point where the principle is abandoned in order to maintain the system, in a pattern of organised deceit designed to reproduce the interests of a particular class. The victory of liberalism over all other world views has closed down our political horizons, for once an ideology quashes the opposition, it becomes increasingly dogmatic (Feyerabend, 1978). Hence the existing climate consensus is inherently reactionary, an ideological support structure for securing the socio-political status quo (Swyngedouw, 2022).

The future is always plagued with uncertainty and there is no reason to suppose that our climate future should be any more free of uncertainty than any other projection. One way in which liberal climate ideology operates is to elide those uncertainties from the way the 1.5°C/net zero future is discussed in politics and campaigns. Uncertainties about the timing, extent and severity of climate impacts persist and are perhaps irreducible (e.g. Howe et al., 2019). Uncertainties also persist around what cuts in emissions are needed to limit warming to 1.5°C, with climate policies offering only a 66% chance of limiting warming to 1.5°C. These uncertainties have been described as possessing the characteristics of a 'monster' (van der Sluijs, 2005). The uncertainties are monstrous because they have escaped the categories that liberalism has sought to impose on the world. Liberalism would like to keep economics, politics, society and science as separate domains (Mann, 2018). Climate change pays no heed to distinctions between categories such as facts versus values and science versus policy (van der Sluijs, 2005).

A crucial area of uncertainty is about what happens to global temperatures if and when the world reaches net zero. Such uncertainty – though an inevitable feature of scientific projections of future behaviours of complex systems – again underlines the ideological nature of simplistic and performative claims that 1.5°C of warming is a solution to climate change. One recent analysis of a number of different models showed a wide range of possible outcomes, from ongoing warming for thousands of years after we have reached net zero to, at the other end of the scale, substantial cooling. The authors concluded that overall, the most likely outcome is that the world will stay at the 1.5°C threshold of warming for many decades (MacDougall, Frölicher and Jones, 2020). In other words, existing climate policies at best will leave humanity at the precipice of disaster for decades. It bears repeating – that is the best-case scenario, and is itself based on questionable assumptions about the rates of change, how the climate and biosphere will respond to warming, technological progress and the functioning of society under extreme environmental duress (Hausfather, 2018).

The scientific uncertainties regarding how to reach a climate-safe future are compounded by uncertainties about emission reductions needed for 1.5°C. The authoritative online climate news service Carbon Brief has highlighted the ambiguity of these numbers, which hide huge complexities and offer varying carbon budgets, depending on the methodologies and assumptions used to calculate the figures (Hausfather, 2018). The Carbon Brief analysis reports that, based on estimates made in the IPCC's fifth assessment report (AR5), there would be around 120 gigatonnes of CO_2 (GtCO_2) remaining from the beginning of 2018 – or around three years of current emissions – for a 66% chance of avoiding 1.5°C warming. The IPCC's Special Report on the emission cuts needed to limit warming to 1.5°C, released in 2018, raises the budget for a 66% chance of avoiding 1.5°C to 420GtCO_2 – or 10 years of current emissions (IPCC, 2018). It does this by revising past figures and claiming that previous models underestimated how much CO_2 was emitted in the past. Raising the amount that was emitted in the past leads to the conclusion that the climate is not as sensitive as previously thought. It took more CO_2 than thought to raise the temperatures to current levels, so we can emit more CO_2 in the future than we thought before we hit 1.5°C.

However, Carbon Brief go on to report that the larger remaining budget for greenhouse gas emissions is only possible through the large-scale use of negative emissions in the future and assuming the world goes beyond 1.5°C of warming before than coming back to 1.5°C in the future by sucking excess amounts of CO_2 out of the atmosphere and storing the CO_2 underground permanently. Unfortunately, the validity of this assumption is looking increasingly shaky. Wildfires have meant forests planted to offset emissions in California have actually emitted between 5.7 and 6.8 million tonnes of carbon since 2015 (Hodgson, 2022). Meanwhile, a recent report has concluded carbon capture and storage schemes are not working and cannot offer any solution to climate change (Gayle, 2022). The European Union (EU) is spending at least 50 billion Euros in 2022 on new and expanded fossil fuel infrastructure, 'threatening to put emission targets at risk' (Hancock, 2022). At the same time, a key plank of the EU's emission reductions strategy – the importation of wood pellets from overseas forests to be burnt in Europe's power stations – is coming under increasing attack as an energy strategy that will produce higher emissions than fossil fuels over the coming decades (Thunberg, 2022).

Voices from the global South, and subaltern voices from the North, have been largely excluded from the process of defining acceptable climate risk under conditions of uncertainty. We would anticipate such exclusions, given that liberalism has little to say on the issue of inequality. Yet, as the growing climate justice movement attests, there is an increasing recognition of the role inequality plays in stopping action on climate change (Ehrhardt-Martinez et al., 2015).

Climate justice is a relatively complex matter drawing on philosophy, ethics and political science, with a range of sometimes diverse definitions (Walker, 2012: 39). It has long been known that the impacts of climate change will fall disproportionately upon developing countries and poor persons within all countries, and thereby exacerbate inequities in health status and access to adequate food, clean water and other resources (IPCC, 2001: 12) and global poverty (Harris, 2013: 123). The UNFCCC espouses a declared commitment to justice in both process and outcomes, proposing a hybrid standard by which liability is assigned according to three criteria; equity, historical responsibility and the respective capabilities of nations to reduce their emissions (UNFCCC, 1992; Vanderheden, 2008: 73). Equality, involvement/ participation, fairness, access (to resources and information) and protection from environmental risks are recurrent themes in definitions of climate justice. These themes are very much interrelated, with discussion of one inevitably leading to discussion of another aspect of justice (Walker, 2012: 12). There is a belief running through these discussions that without just processes of deliberation, there can be no just outcomes. Rather than seeing subaltern classes as a threat to the liberal order, it is supposed that including lay perspectives in discussions of how to respond to climate change will socialise the debate and consequently might lead to more socially acceptable, and indeed better, mitigation and adaptation strategies (better because such strategies might be ones that otherwise might not have been thought of by centralised planners) (Lowe et al., 2006). International climate agreements are grounded in a view that sees the world in need of scientific monitoring and management or what Moss called green governmentality (Moss, 1998, cited Bäckstrand and Lövbrand, 2007: 128). Moss describes green governmentality as a global form of power tied to the modern administrative state, mega-science and the business community (ibid.). It is defined by a 'managerialist discourse', which is characterised by an optimism about supposed progress (Broadhead, 2002: 102). This top-down understanding of the world marginalises alternative views (ibid.: 128) and leads to the creation of publicly inaccessible storylines that favour technocratic governance and research elites (Bohemer-Christiansen 2003: 128). Governmentality ensures the security of the state by monitoring, shaping and controlling the people living on that territory (Backstrand and Lovbrand, 2007: 126). The goal is to control and restrict activity, green governmentality extends this to the environment.

Many of the issues of (in)justice are the result of imbalances in the power relations amongst climate change stakeholders, and these imbalances are sustained through language and the way language constructs climate change as a problem solvable through top-down interventions within liberal norms (Shaw, 2013). Harris

(2013) has shown how sustainability discourses, with their emphasis on markets, consensus and ecological modernisation mean questions of equality and procedural rights are too easily downplayed and pushed aside. Additionally, the concentration of wealth into the hands of the lucky few has created an enormous inequality regarding the ability of different groups to participate in public decision-making, allowing the public space to be dominated by powerful and wealthy organisations. Even in spaces where the public are invited to participate in climate decision-making, policy does not emerge as the result of an open public debate to ascertain the common interest. Instead, the public sphere becomes an arena for the sophisticated manipulation of public opinion by the powerful actors. An example of this manipulation is to be found in climate assemblies, where the choice offered to participants is limited to which interventions are necessary to deliver net zero goals. It is the constraining of these discussions within the net zero framing that is doing the ideological work of manipulating the discussion, so as to keep it safely within the climate guardrails being discussed in this book.

Climate injustice is not simply about who is and who isn't invited to the table, but also what forms of knowledge and enquiry are employed. Commitment to a particular form of knowledge predetermines the kind of generalisations one can make about the present world, the kinds of knowledge one can have about it, and hence the kind of projects one can legitimately conceive for changing that present or for maintaining its form indefinitely (Whyte, 1973: 2). (White, male) European ontologies, epistemologies and research programmes have thoroughly conquered the world, suffocating all other approaches (Reiter, 2018: 3). These approaches of Western science do not contain the tools to ask different questions, but as noted above, constrain research and policy development within the framework of liberalism. The universalist claims of liberalism are only made possible by colonialism's obliteration of other forms of knowledge, or what Reiter calls epistemicide (2018: 4). We are left with knowledge produced by researchers from only five countries (Germany, France, Italy, UK and US) (ibid.) or even if conducted overseas, done so in the service of the Western epistemology. An example of the harm done by this liberal climate imperialism is to be found in fire management practices. Crosby (1986) describes how European ideas of fire suppression and fire management have been exported to other climatic regimes with devastating consequences. Australia has a range of species dependent on fires for their regeneration, many forests in the US are meant to burn. Where humans have built settlements in areas meant to burn, they have put in place fire management processes that change the moisture and biochemical resistance in the soil, creating hotter and more intense fires further worsened by climate change.

There is no single 1.5 degree world, except in the policies of liberal institutions. Temperatures are changing at different rates in different parts of the world, natural systems are being impacted in different ways, and the ability of social systems to cope with those impacts varies. Despite the manifold uncertainties surrounding future climate impacts, the idea of diverse individuals with diverse experiences, vulnerabilities, fears and aspirations, each with equal rights are largely absent from these universalising discourses.

3.5 Local experiences of a global phenomenon

The dominant top-down discourse of climate change adopts a technocratic, globalised view in which place-based human scale relationships and experiences are largely absent. But, to be rooted in a place or culture is perhaps the most important and least recognised need of the human soul (Weil, cited Stears: 101). It is for this reason that climate change is always local (Scranton, 2018). In this, it is like all other aspects of human experience, the human scale is local, it is what I can hear, feel and see. But the local can be a very strange and alienated place today in the West.

Alienation from the people around us is sadly all too common. I get on well with my neighbours in the small suburban close that has been my home for 15 years. I know most of their names, and I know, at a very superficial level, what some of them do for a living. I sometimes socialise with some of them, they are good people. But I don't really know what they get up to during the day, how they ensure the conditions for their survival and reproduction of their labour. Unless I am socialising with them, I don't know what any of them are doing of a weekday evening or weekend. I really know nothing about their day-to-day existence, how they spend their time, what they are thinking and feeling. When we meet up, we will recount at a surface level some of the things that have happened, steering clear of issues where there may be a difference of opinion.

This situation would have been unthinkable for most of human existence. Imagine living as a group, a nomadic tribe or small settlement, and not knowing what the people who lived around you did during the day, having no shared skills or common knowledge or views about the world. Imagine not knowing what the people next to you do at night, how they keep themselves alive, what has happened to them that day. Or children not even really knowing what their parents do during the day, how they provide the sustenance you need, where they go, who they go there with. This is a profoundly strange and historically unique situation to find ourselves in.

For Enlightenment thinkers – Locke, Rousseau and Kant – liberalism is the means by which we might achieve identification of the one with the all, with all humanity and each human being. Yet the starting place today for such universal identification, in many parts of the West at least, is a world in which many people cannot even identify with the person who lives 20 metres away. The only way to achieve that identification is to make the rest of the world like the West.

Reflecting on the loss of our Enlightenment ideals, the cultural critic de Zengotita mourns that most people do not see the world through the cosmopolitanism of liberal ideals. They don't care and are not interested in identifying with people who have very different lives from them. One reason for this is because the ability to think in the abstract is not a universal trait amongst humans (de Zengotita, 2003). It is a way of thinking limited to the well-educated middle class. And it is the middle class who are most likely to be living somewhere where local and community ties have the least meaning, making it easier to imagine the local as not particularly relevant to discussions of climate change. In 'The New Class War' (2020) Michael

Lind reflects on David Goodhart's comparison of the 'communitarian localism' of the "somewheres" with the individualistic careerism of educated "anywheres" (Lind, 2020: 24). For many working class "somewheres", personal identities as members of particular local communities or extended families are more important than their low-status jobs. In contrast "anywheres" prioritise high-status jobs at the expense of ties to any particular place or community (ibid.). And for all the fetishisation of "communities" in liberal discourses, what gets between us and other people are the anxieties brought about by social inequality, an issue on which liberalism remains mute (Wilson and Pickett, 2018).

Lind compares the global elite institutions that are designing our lives with the cross-class institutions that were familiar in the post WWII settlements of the West and in which the working class had a voice; mass membership political parties, unions, grassroot religious and civic institutions (Lind: Lind, 2020)). What Lind describes as the intentional destruction of these institutions has robbed the working class of agency (2020: 88). Liberal run social justice and climate change NGOs have nothing to offer the working class today, they have nothing to offer but the entrenchment and reproduction of liberalism (Lind, 2020: 144). The construction of climate change in global terms rather than local is another arena in which the working class have lost agency.

3.6   Conclusion

Climate change policy is characterised by the universalisation of liberal ontologies. The local is sacrificed to the global, the particularities of place, the need to be rooted, is supplanted by a vision of the good life as one of having the whole planet as one's home. We are encouraged to sever the ties that leave us trapped in a community, and travel the world in search of money and self-actualisation. This idea of how to live is specific to a particular geographical, historical, and class-based experience of the world. Climate policy aligned to liberal Western interests has been globalised through an institutional network. In the early 1990s, these institutions succeeded in identifying climate change as an issue of sustainable development, for example, sustaining economic growth in the face of environmental challenges. The scientific uncertainties inherent in trying to predict future climate change trajectories and impacts were overcome by the development of a single global dangerous limit, set at a level that it was thought would be manageable in wealthy Western countries, and building international agreements around this single global target.

References

Avigur-Eshel, A. (2014). 'The ideological foundations of neoliberalism's political stability: an Israeli case study'. *Journal of Political Ideologies*, Vol. 19, No. 2, 164–186. doi: 10.1080/13569317.2014.909261

Avigur-Eshel, A. (2019). 'Speaking to the outraged: discursive responses of policy elites to social unrest over economic issues'. *Critical Policy Studies*. doi: 10.1080/19460171.2018.1463857

Bäckstrand, K. and Lövbrand, E. (2007). 'Climate governance beyond 2012: Competing discourses of green governmentality, ecological modernization and civic environmentalism', In Pettenger, M. (ed). *The Social Construction of Climate Change. Power, Knowledge, Norms, Discourses*. London: Routledge, pp. 123–148.

Bernstein, S. (2002). *The Compromise of Liberal Environmentalism*. New York: Columbia University Press.

Betty, L (2017). *Without God: Michel Houellebecq and Materialist Horror*. Pennsylvania: Pennsylvania State University Press.

Boehmer-Christiansen, S. (1994). 'Global Climate Protection Policy: The Limits of Scientific Advice'. *Global Environmental Change*, Vol. 4, No. 2, 140–159.

Broadhead, L. (2002). *International Environmental Politics–The Limits of Green Diplomacy*. Boulder, CO: Lynne Rienner Publishers.

Crosby, A. (1986). *Ecological Imperialism: The Biological Expansion of Europe, 900–1900*. Cambridge, MA: Cambridge University Press.

de Zengotita, T. (2003). 'Common Ground: Finding Our Way Back to the Enlightenment.' *Harper's Magazine*. Retrieved from https://www.thefreelibrary.com/Common+ground%3a+finding+our+way+back+to+the+enlightenment.+.-a099601601

Dunlap, R. and McCright, A. (2015). Challenging climate change. The denial countermovement'. In Dunlap, R. and Brulle, R. (eds). *Climate Change and Society*. Oxford: Oxford University Press, pp. 300–332.

Eden, L. (2004). *Whole World on Fire. Organizations, Knowledge and Nuclear Weapons Devastation*. New York: Cornell University Press.

Ehrhardt-Martinez, K., Schor, J., et al. (2015). 'Consumption and climate change'. In Dunlap, R. and Brulle, R. (eds). *Climate Change and Society*. Oxford: Oxford University Press, pp. 93–126.

Feyerabend, P. (1978). *Science in a Free Society*. London: Verso.

Gayle, D. (2022). *The Guardian: Carbon Capture Is Not a Solution to Net Zero Emissions Plans, Report Says*. Retrieved from https://www.theguardian.com/environment/2022/sep/01/carbon-capture-is-not-a-solution-to-net-zero-emissions-plans-report-says

Gordon, L. (2015). *What Fanon Said. A Philosophical Introduction to his Life and Thought*. London: Hurst and Company.

Greif, M. (2015). *The Age of the Crisis of Man: Thought and Fiction in America, 1933–1973*. Princeton, NJ: Princeton University Press.

Hamilton, G. (2009). 'Narrative Policy Analysis and the integration of public involvement in decision making'. *Policy Sciences*, Vol. 42, 227–242.

Hancock, A. (2022). *Financial Times: Europe's New Dirty Energy: The Unavoidable Evil of Wartime Fossil Fuels*. Retrieved from https://www.ft.com/content/b209933f-df7f-49ae-8f82-edc32ed622a6

Harris, P. (2013). *What's Wrong with Climate Politics, and How to Fix It*. Cambridge: Polity Press.

Hausfather, Z. (2018). 'Analysis: Why the IPCC 1.5C report expanded the carbon budget'. *Carbon Brief*. Retrieved from https://www.carbonbrief.org/analysis-why-the-ipcc-1-5c-report-expanded-the-carbon-budget/

Hodgson, C. (2022). *Financial Times: Wildfires Destroy Almost All Forest Carbon Offsets in 100-Year Reserve, Study Says*. Retrieved from https://www.ft.com/content/d54d5526-6f56-4c01-8207-7fa7e532fa09

Howe, L.C., MacInnis, B., Krosnick, J.A., et al. (2019). 'Acknowledging uncertainty impacts public acceptance of climate scientists' predictions'. *Nature Climate Change*, Vol. 9, 863–867.

IPCC. (2001). *Climate Change 2001: The Scientific Basis. Contribution of Working Group I to the Third Assessment Report of the Intergovernmental Panel on Climate Change*. Houghton, J.T., Y. Ding, D.J. Griggs, M. et al. (eds). Cambridge, UK and New York, NY, USA: Cambridge University Press, 881pp.

IPCC. (2018). *Summary for Policymakers. In: Global Warming of 1.5°C. An IPCC Special Report on the Impacts of Global Warming of 1.5°C Above Pre-Industrial Levels and Related Global Greenhouse Gas Emission Pathways, in the Context of Strengthening the Global Response to the Threat of Climate Change, Sustainable Development, and Efforts to Eradicate Poverty* [Masson-Delmotte, V., P. Zhai, H.-O. Pörtner, D. Roberts, J. Skea, P.R. Shukla, A. Pirani, W. Moufouma-Okia, C. Péan, R. Pidcock, S. Connors, J.B.R. Matthews, Y. Chen, X. Zhou, M.I. Gomis, E. Lonnoy, T. Maycock, M. Tignor, and T. Waterfield (eds.)]. Cambridge, UK and New York, NY, USA: Cambridge University Press, pp. 3–24.

Johnson, B. and Covello, V. (eds). (1987). *The Social and Cultural Construction of Risk*. Dordrecht: Kluwer Academic Publishing.

Lähde, V. (2006). 'Gardens, climate changes and cultures: an exploration into the historical nature of environmental problems', in Haila, Y. and Dyke, C. (eds). *How Nature Speaks. The Dynamics of the Human Ecological Condition*. Durham, NC: Duke University Press, pp. 78–105.

Lind, M. (2020). *The New Class War: Saving Democracy from the Metropolitan Elite*. London: Atlantic Books.

Liverman, D.M. (2009). 'Conventions of climate change: Constructions of danger and the dispossession of the atmosphere'. *Journal of Historical Geography*, Vol. 35, No. 2, 279–296.

Lowe, T., Brown, K., Dessai, S., Franca Doria, M., Hayes, K. and Vincent, K. (2006). Does tomorrow ever come? Disaster narrative and public perceptions of climate change. *Public Understanding of Science*, Vol. 15, 435–457.

Lukes, S. (2017) *Liberals and Cannibals: The Implications of Diversity*. London: Verso.

MacDougall, A.H., Frölicher, T.L., Jones, C.D., et al. (2020). 'Is there warming in the pipeline? A multi-model analysis of the Zero Emissions Commitment from CO2'. *Biogeosciences*, Vol. 17, 2987–3016.

Machin, A. (2013). *Negotiating Climate Change: Radical Democracy and the Illusion of Consensus*. London: Zed Books.

Mair, P. (2014). *On Parties, Party Systems and Democracy: Selected Writings of Peter Mair*. Colchester: ECPR Press.

Mann, G. (2018). *In the Long Run We Are All Dead*. London: Verso.

McCright, A. and Dunlap, R (2011). 'The politicization of climate change and polarisation in the public's view of global warming 2001–2010'. *The Sociological Quarterly*, Vol. 52, 155–194.

O'Riordan, T. and Jordan, A. (1999). 'Institutions, climate change and cultural theory: Towards a common analytical framework'. *Global Environmental Change*, Vol. 9, No. 2, 81–94.

Pettenger, M.E. (2007). *The Social Construction of Climate Change. Power, Knowledge, Norms, Discourses*. London: Routledge.

Piccolo, J., Taylor, B., Washington, H., et al. (2022). 'Nature's contributions to people and peoples' moral obligations to nature'. *Biological Conservation*, Vol. 270.

Porter, G. and Brown, J.W. (1991). *Global Environmental Politics*. Oxford: WestView Press.

Reiter, B. (2018). *Constructing the Pluriverse. The Geopolitics of Knowledge*. London: Dukes University Press.

Schmidt, V. (2008). 'Discursive Institutionalism: The Explanatory Power of Ideas and Discourse.' *Annual Review Political Science*, Vol. 11, 303–326.

Schmidt, V. and Radaelli, C. (2004). 'Policy change and discourse in Europe: Conceptual and methodological issues'. *West European Politics*, Vol. 27, No. 2, 183–210.

Scranton, R. (2018). *We are Doomed, So Now What?* London: Soho Press.

Shaw, C. (2011). *Choosing a Dangerous Limit for Climate Change: An Investigation into How the Decision Making Process Is Constructed in Public Discourses*. DPhil. Brighton: University of Sussex.

Shaw, C. (2013). 'Choosing a dangerous limit for climate change: Public representations of the decision making process'. *Global Environmental Change*, Vol. 23, No. 2, 563–571.

Stipple, J. and Paterson, M. (2007). 'Singing Climate change into existence: on the territorialisation of climate policymaking', in Pettenger, M. (ed). *The Social Construction of Climate Change: Power, Knowledge, Norms, Discourses (Global Environmental Governance)*. Aldershot: Ashgate Publishing Limited, Pp. 147–171.

Streeck, W. (2020). Fighting the state', in *Critical Encounters. Democracy, Capitalism, Ideas*. London: Verso, Pp 113–130.

Swyngedouw, E. (2022). 'The unbearable lightness of climate populism'. *Environmental Politics*, Vol. 31, No. 5, 904–925.

Thunberg, G. (2022). *The Guardian: Burning Forests for Energy Isn't 'Renewable' – Now the EU must Admit It*. Retrieved from https://www.theguardian.com/world/commentisfree/2022/sep/05/burning-forests-energy-renewable-eu-wood-climate

Thwaites, J. (2022). David Malpass is a climate and development failure. *NRDC.org*. Retrieved from https://www.nrdc.org/experts/joe-thwaites/david-malpass-climate-and-development-failure

UNFCCC. (1992). *United Nations Framework Convention on Climate Change*. Retrieved from http://unfccc.int/resource/docs/convkp/conveng.pdf.

Van der Sluijs, J. (2005). 'Uncertainty as a monster in the science–policy interface: Four coping strategies'. *Water, Science and Technology*, Vol. 52, No. 6: 87–92.

Vanderheden, S. (2008). *Atmospheric Justice: A Political Theory of Climate Change*. London: Oxford University Press.

Walker, G. (2012). *Environmental Justice. Concepts, Evidence and Politics*. London: Routledge.

Whyte, H. (1973). *Metahistory: The Historical Imagination in Nineteenth Century Europe*. Baltimore, MD: John Hopkins University Press.

Willis, R. (2017). 'Taming the climate? Corpus analysis of politicians' speech on climate change'. *Environmental Politics*, Vol. 26, No. 2, 212–231.

Wilson, R. and Pickett, K. (2018). *The Inner Level: How More Equal Societies Reduce Stress, Restore Sanity and Improve Everyone's Well-being*. London: Allen Lane

Wynne, B. and Jasanoff, F. (1998). *Science and Decision Making in, Human Choices and Climate Change, Vol. 1. The Societal Framework*. Rayner, S. and Malone, E. (eds). Ohio: Battelle Press.

4 Guardrail 3: Climate change is not an historical phenomenon

Historical change and social conflict are irrelevant to solving climate change. Climate change is not a driver of historical change in the way that material and dialectic forces have driven historical epochs in the past. Climate change can be solved within the existing relations of production that characterise liberal democracies. Climate change should be de-politicised, and responded to purely on the basis of the science and agreed targets.

An event or process is historical when it utterly and irrevocably changes how people see and act in the world. Humans have experienced few such changes. The move into historical time is one such event, associated with the advent of agriculture and cities. The shift from feudalism to industrial capitalism is another.

Ulrich Beck in his proposal of an emerging 'risk society' (Beck, 1992) anticipated that awareness of the environmental risks posed by modernity would lead to a transformation of society, a new 'reflexive modernity'. A reflexive modernity emerges as capitalist modernisation begins to undermine the foundations for its own existence. Under these conditions, the legitimacy of modernity itself comes increasingly under attack from its own citizens (Adam, Beck and van Loon, 2000). The 'reflexive modernity' as described by Beck has yet to emerge, despite widespread recognition of the damage being wreaked on the planet. A recent global survey of 110 countries found majorities in 108 of the 110 countries reported feeling either "very" or "somewhat" worried about climate change (Leiserowitz et al., 2022). This widespread anxiety has not 'altered or destroyed the foundations of the societal order' (Adam, Beck and van Loon, 2000: 114). Nor has climate awareness, as Adam et al. claimed 13 years ago, begun 'transforming the political, economic, socio-technical, legal, military and cultural landscape' (ibid: 114–115). Instead, it is becoming increasingly apparent that liberalism represents the end of the line historically; liberalism's victory over history is complete (Noble, 2012). This is as good as it gets and now all that remains is to make everyone liberal, and stay liberal.[1]

It is difficult to imagine a different society from this one in part because our notions of time and history are tied up with technological rather than social change; we mistake technological change for historical change. Material progress confers modernity with this otherwise absent sense of purpose, making increased

DOI: 10.4324/9780429463488-5

consumption both a moral imperative and humanity's necessary and inevitable destiny. As a result, time is no longer understood as a circular phenomenon – the typical perception of non-industrial societies – but is instead linear and irreversible, as expressed through the infinite accumulation of knowledge. Wealth and knowledge are both forms of capital, and it is this marking of time through the accumulation of these two forms of capital which generates a kind of philosophical force, pushing us ever 'forward' and blocking the path behind us. The sum of these forces, narratives and promises forecloses any sense of being able to 'go back and we are instead imperilled to chase after the tomorrow which is always better than today' (Dean and Zamora, 2021: 8). It has been suggested our environmental ills can be addressed by going back in time. What perhaps is so transgressive about such an idea is not so much returning to a previous way of life, but, – and this really is the critical point – having returned there, then stay there. To say goodbye to a world defined by ceaseless technological change is truly preposterous, utterly unimaginable.

Liberalism should be an outlook, a standpoint, a perspective. Instead, it has become dogma. So, we find ourselves, at the apogee of liberalism, as free as we could possibly hope to be, facing catastrophe, and there is no vibrant discourse in the workplaces, homes, barracks and prisons of the country about alternative political, economic and ideological ways of being in the world. The tenets of reason and rationality upon which liberalism has been built would suggest that, confronted with humanity's greatest historical challenge, a number of different political philosophies would be explored as a part of the rational process of democratically deliberating on how to respond. Yet within the majority of academic, environmental, cultural and political debate the necessity, indeed the possibility, of looking outside of liberalism no longer arises.

Being unable to escape the problems of modernity by returning to the past, we can only look to the structures and discourses of modernity for a solution. The idea of revolution that dominated European thought since the French revolution of 1789 has disappeared. 'We no longer seek to change the world, we are content to simply change ourselves' (Michaels, 2004, cited Dean and Zamora, 2021: 8). Liberalism has always been the instrument of a particular class in social evolution, the capitalist class. A new opposition emerged between wage earners and the capitalist class in the early days of industrialism and 'liberalism changed from being an ideology of radical change and progress into a defence mechanism the capitalist class deployed against the workers, becoming an ideology of reaction and conservatism' (Misra, cited Laski, 1936: 10). Except it seems circumstances have caught up with us and historical change may be inescapable, with growing calls for a research programme predicated on the assumption that humanity is now entering the climate endgame and societal collapse (Kemp et al., 2022).

4.1   De-historicising the transformation

We are living through the deep time equivalent of an asteroid strike, so quickly have we released such vast quantities of stored solar energy into the atmosphere.

A radical approach to the crisis, according to Parenti, begins with the recognition that we have very little time to respond to climate change. These urgent deadlines mean, there is not enough time to try and bring about historical change (Parenti, 2013).

This may be true, and acceptance of the fact that we no longer have time to pursue historical change is to confirm liberalisms' final victory over history. Holloway argues our only way out of our current predicament is through historical change but to adopt a properly historical perspective would be suicidal for liberalism because it would reveal the transitory nature of its own hegemony (Holloway, 2019: 81). The impossibility of liberalism allowing the exploration of post-liberal climate futures exposes a profound conflict at the heart of liberalism. Liberalism was supposed to offer the chance to break free of hidebound thought, to replace custom with experiment. Dissensus, in the form of open public debate and the clash of value positions is what liberal cultures are supposed to promote and sustain (Lukes, 2017: 39).

In times past, the streets and our culture would be filled with arguments about our possible different political futures. Culture is a vital arena for exploration of social transformation. Culture is, after all, what people are willing to fight and kill for – language, belief, kinship, heritage and homeland (Eagleton, 2016: 128). White identifies entertainment, academic orthodoxy and political ideology as the three great obstacles to our ability to imagine a different future (2003: 6–7). These three themes – under the label of culture – have as their principal task not the catalysing of social transformation, but instead the maintenance of behavioural stability by circumventing the use of force, via seduction and intimidation (Peckham, 1995: xvi). Culture, in other words, controls behaviour (ibid.).

If transformation does not mean historical change, what does it mean? Is the word being used in a reactionary way, to capture the imaginative terrain of transformation and populate it with non-transformative actions? Transformations at a general level are understood as fundamental changes in and across various domains spanning from individuals' mindsets, attitudes, and beliefs to social norms and practices, to institutions and political systems (Brown et al., 2013; O'Brien and Sygna, 2013). Such definitions would seem to demand that the transformation needed is historical in nature, and implicates all people in the process, if mindsets, attitudes and political systems all have to be transformed.

4.2 Removing the working class from the transformation

Transformations for environmental or climatic sustainability are seen to require mobilisation of actors in collaborative efforts (Huitema and Meijerink, 2010). Exactly how these shared understandings and ideas for interventions emerge in collaborative networks is still an area of research in transformations studies (Moore et al., 2018). There are immense challenges involved in recruiting people to collaborate in bringing about fundamental change. Orwell highlighted some of those challenges with his eye on a separate kind of social transformation, that needed to bring about socialism. In *The Road to Wigan Pier*, Orwell argues that ordinary people ("the common man") do not realise the immense changes needed for socialism (or in our case, needed to limit the harms caused by climate change). Their vision of socialism (or for our purpose, a net zero future) is the present with the worst abuses taken out. 'We live in the wreckage of a civilisation but it was a great civilisation and it still has its bouquet, here and there' (our freedoms, consumption, holidays, the promise that tomorrow could be different) whereas 'the socialist future tastes only of iron and water' (the net zero future just looks the same as today but with the treasured freedoms removed) (Orwell, 2001: 170). Orwell's insights also reveal another challenge in trying to bring about historical change on a collective basis (e.g. by including those – perhaps the majority of humans – for whom thinking about an alternative future society is not a familiar activity). What Orwell saw was that people will only use up their lives in heartbreaking political struggle not in order to establish some 'central-heated air-conditioned strip-lighted paradise' (or in our case, a heat source pump paradise) but 'because they want a world in which human beings love each other' (Orwell, 1943).

Orwell went on to remark that grand vaulting abstractions about historical change feature no clear and concrete responses to the everyday problems people face (ibid.). As a result, the discussions of historical change are colonised by a particular set of voices, the voices of those with the education that makes it easier to deal with the abstract reflections of humanity's future. This class of professionals and highly educated are the most ideologically and politically conservative of all sectors of society (Schmidt, 2001). The philosopher Groys explores Derrida's claim that fear of an apocalypse such as nuclear war is not simply fear of death – we are going to die anyway, nuclear war or not. Rather what is feared is the destruction of 'museums, libraries, and all depositories of created works, everything written, painted and so on, in which the intellectuals of today, not believing in any transcendence, seek social and historical immortality within the world' (Groys, 2102: 83). It is the loss of all this which threatens the secular intellectual's immortality, the end of a society that can sustain a process of cultural production which embodies middle class conceptions of what we mean when we talk about civilisation. The middling classes are the ones who, by going to the museums, reading the books and visiting the theatres, give legitimacy to these works, and who are the carriers of the creator's immortality. Their own identity as middle class is inextricably tied up with their cultural choices. They are co-producers of this civilisation, embodied in these art works, and so the continuity of their way of life, of civilisation, requires the existence of these repositories, and an ongoing parade of people like themselves to stand or sit before these works of art and say to themselves and each other "this is what it means to be human, this is what it means to be civilised".

It does not take a great leap of the imagination to substitute fear of the nuclear apocalypse with fear of catastrophic climate change. Our ability to produce and sustain a middle class culture might well be at stake if some of the climate change projections were to come true. But what have the working class got to fear from such a calamity? They have no presence in the cultural mausoleums of our age, there is no immortality for them, and so no apocalypse. The most they can hope for is to have their name on a small brass plaque fixed to a bench in a park. So, it is this very construction of what is at stake from an issue such as climate change, what it is deemed worth preserving, that excludes the working class. In his book on the psychology of climate change, the economist Per Epsne Stokes argues that from a cognitive psychology point of view, the main reason why so few individuals take to the street to protest about climate change is because the whole issue is presented as very distant to us (Stoknes, 2015: 39). From a political and ideological point of view, one might argue the reason so few people take to the street to protest is because the kind of goals people are protesting in support of mean nothing to the vast majority of the working class.

Hume thought humanity is much the same in all time and places, in so much as people lack a common nature and are instead creatures of circumstances. If those circumstances are benign, familiar and suited to a natural way of living then humans will shuffle along without being possessed with a need to change or trans-form the world. In a similar vein, Lukes maintains 'what is universal in humans is not actually realised behaviour but potentialities and capacities and tendencies and vulnerabilities' (2017: 13). Under liberalism everyone is free to be an individual to express that individuality, as long as they act as an individual, that is, in their behaviour they reproduce the ideology of liberalism. 'History, for the bourgeoisie, is the eternal task of people returning to the arena where they could display their individuality' (Whyte, 1973: 423).

There is a large disconnect between what liberal political discourses say is required (inconvenient changes to the way we source and use energy, in an other-wise unchanged set of political and economic circumstances), what the public feel is a sufficient and achievable amount of behaviour change (recycling, reduction of meat consumption, turning down radiators) and the fundamental and historical transformation of our world identified as necessary by climate scientists.

4.3   Intellectuals and the de-historicising of climate change

There is no desire for truly historical change within the political institutions of the West. Meaningful contact with the government and policy is only available to insider groups, those with shared values. Roberts (2004: 148) describes an elite model of how governments function in liberal democracies. In this model, the government is a servant of capital and the only ones who need to raise issues are the powerless. This is because elite interests are always catered for, and are embedded into the system. The political cues born of these elite interests are the biggest influence on public attitudes towards climate change – the political consensus drives a public consensus (Brulee et al., 2012). On any issue where political elites agree there is no political contestation, the issue does not even appear on the political agenda.

It is a truism within liberal climate change discourse that greater public participation in decision-making is to be encouraged, and that such participa-tion will lead to better climate change decision-making. Too often participation is an exercise in building consent for liberal visions of humanity's future and the marginalisation of alternative visions of humanity's future. The actual trajectory and principles of liberal capitalist 'technological society' (Ellul, 1965) are not part of the participatory project. Whilst there is a significant body of research into the weakness of the participatory agenda, such critiques are largely absent from public discourses of the need to build greater public engagement with climate change policy. Where criticisms appear, they are used to argue for a more refined form of participatory activity which shares the same goal as the more explicitly manipulative end of 'manufacturing consent' spectrum, namely management of resistance to the risks generated by liberal philosophies of human progress. The shift towards consensus lies with the public, whilst the experts stay where they are (Habermas, 1998). 'The intellectuals have succeeded in preventing a more direct democracy where problems are solved and solutions judged by those who suffer from the problems and have to live with the solutions' (Feyerabend, 1978: 85–86).

Marx and Engels described how 'modern bourgeois society is like the sorcerer who is no longer able to control the powers of the nether world who he has called up by his spells' (Marx and Engels, 1888: 58). In order to win the moral consent of the population for life in an out-of-control, world destroying economic system, liberalism must propose a set of descriptions of the world, and the values that preside over it, to become internalised by the public. The intellectual strata and civil society associations are two of the actors key to this. Another is journalism. An important function of journalism is to maintain the population in a constant state of desire which people attempt to assuage through consumption. Apocalyptic headlines grab the audience's attention and hence ensure visibility for the advertising and inducement to consumption which accompany the stories. The apocalyptic scenarios in the headlines are resolved in the story through narratives of technological solutions, global negotiations and occult economic practices. The effect is to assimilate climate change within the framework of liberal imaginaries whilst depoliticising the issue. It has been noted that the news media is the primary means by which people encounter climate change (Doulton and Brown 2009: 199). So, what does this mean for public attitudes to climate change? It might be tempting to assume (as much commentary does) that what is reported by the official 'organs of opinion' (Gramsci, cited Anderson, 2017: 20) maps on directly to what people think and do. Political language, whether used by decision-makers, the media or environmental NGOs, works as a purposeful and directed 'linguistic action' (Wodak, 2008: 5) that seeks to act upon the social world in significant ways. News reporting and commentary – as part of the military-industrial-media complex (Ross, 1991: 6) – presents a version of reality aligned broadly with the norms of bourgeois liberal democracy.

If the political is related to the social, the collective, then the focus on the individual as the agent of transformation is inherently political, in the sense of downplaying the role of political action for historical change in favour of individual choice. There has been growing attention paid to the depoliticisation of the climate change debate (e.g. Carvalho et al., 2017; Hammond, 2018; Malm, 2018). We can best understand efforts to depoliticise and dehistoricise the issue, to imagine that climate change can be solved by bourgeois liberalism, as part of a strategy of control. Liberal control of the narrative is an increasingly important part of liberal hegemony given that the perceived alternative is the mindless and gullible mob being manipulated by right wing populists.

Freedom of thought would pay close attention to the possibility that the liberal climate "solutions" are dead ends. But the apocalyptic scenarios that might arise from failed climate policies are suppressed by the liberal imagination as incompatible with higher rational thought and planning (Sebald, 2014, but see also Ghosh, 2016). W.G. Sebald, in a series of essays on writers, reflects on an almanack that used to be in his grandfather's house in Austria. These almanacks provided ordered accounts of the years' events and rituals of the region where his grandfather lived. The view of the world offered by these almanacks promised there would be no deceit and no violence, and everywhere peace and satisfaction would reign, if, in accordance with the natural order, people were content to provide for themselves with the labours of their own hands. Sebald remarks, 'In such nostalgic utopian views the liberal bourgeois was wont to articulate its discomfort at the rapid spread of the economy of goods and capital, which it itself had created' (2014: 23). This confusion, contradiction and desire to find some middle ground between preserving the benefits of a limitless anarchic market system, whilst maintaining order in the world, finds expression in progressive liberal climate policy. That is why liberalism when confronted by climate change, takes on the mantle of ecomodernism, which shares with modernism a denial of natural limits to capital accumulation and defers the need to confront the possibility of historical change (Foster, 2017).

Many social scientists, as members of the liberal bourgeois, also struggle to entertain thoughts of chaos, collapse and the need for historical change. Instead, they tend to favour scenarios which promise the conditions necessary for the replication of their class's privileges and status. This situation is worthy of explanation, given that since social change constitutes core business for the social sciences, one might expect these disciplines to be taking centre stage, generating lively popular and policy debate about what such transformation might entail and how it might come about (Shove, 2010: 1273). Social scientists, argues Fischer, have pursued the development of a functional relationship to those in power (1987: 101). In this arrangement, the social scientist supplies technical knowledge about the efficient achievement of goals deemed necessary by these elite political actors. Only these people are viewed to be in the position to judge which of the social sciences recommendations can be used when and how (Fischer, 1987: 101). From the very beginning of Western rationalism, 'intellectuals have regarded themselves as teachers, the world as a school, and people as obedient pupils' (Feyerabend, 1978: 121). Social scientists have an important role as "norm entrepreneurs," 'mobilising support for particular ways of talking about and responding to social issues' (Ingebritsen, 2002: 12) and so providing a means of supervisory control (Foucault, 2020: 59). The social sciences construct norms with the aim of judging, correcting and controlling people (Cook, 2018: 84), adopting a role previously the job of priests and pastors (ibid: 87). All the world amounts to for the social sciences is a universe of ideally functioning units, and their role is to describe and fix any deviation from this ideal state (Stears, 2021: 16). Positions in the social sciences are earnt and the funding secured because the research assumptions and goals are aligned with existing social norms (Bernstein, 2002: 20). Liberal environmentalism shapes the research agenda, rather than liberal environmentalism emerging as a response to the research (ibid). Certain kinds of knowledge and policy responses are granted legitimacy not because of their inherent truth or effectiveness but because liberal environmentalism grants them their legitimacy. The goal is to produce results that are believable to others looking over the same events (Bernstein, 2002: 25). Four major problems with social science climate research are apparent; individualisation of the research object; aggregation of individuals into types which avoid the subject of class; constraining future pathways into forms of low-carbon consumption; premising the value of research conclusions on fallacious (overly optimistic) climate change scenarios. It is a common lament within the social sciences that despite the repeated warnings, and much encouragement, sustained and substantive changes in the behaviours of individual citizens have not been forthcoming. Rather than look to corporations, the political system or liberal economics for change, what is sought instead is a measurable increase in reported levels of public concern in response to particular messages, or even an actual change in behaviour such as putting lids on saucepans when boiling water.

This situation is a structural one, these shortcomings are the ones required of the role, and are not the consequence of individual choices. From a Marxist materialist standpoint, the university is an important element of the superstructure. If the production relations in society are capitalist, then the superstructure of action and belief is also capitalist, and the university would obviously serve the interests of the capitalist or bourgeois class (Rodney, 2018: 15). In an article exploring whether the role of the social sciences could ever be otherwise, Amy shows that all analytic methods look at the world in certain limited ways, including some parts and excluding others. These analytical biases 'reflect what issues are considered worth analysing, what facts important to look at, what the public good consists of and so on' (Amy, 1984: 49). Ethical analysis is absent from the work of the social sciences, because ethics are anti-technocratic (ibid.). There is no room for looking beyond the technocratic horizon because the dominant public justification for policy decisions remains technocratic. It is no coincidence that the future we are headed to under liberalism has been given the title the Singularity (though as always with liberalism there is an unresolved contradiction here because under the singularity, we all become one, humans and machines merged in perfect, timeless unity).

4.4   Living with the past

The Returned is a French television drama featuring a remote French town where, one day, the dead start to return. The dead in this programme are not zombies, but apparently normal people who seem as puzzled as the living to find themselves back in the world. The town's inhabitants, both live and returned, are all trying, and failing, to make sense of what is happening. This collective bewilderment is given dramatic momentum by the resentment of the living towards the returned. Asides from intermittent episodes of violence, both physical and emotional, not much happens. People attempting to leave by driving on the only road out of town always find themselves back where they started.

The story can be understood as a metaphor for humanity's current ecological and social problems. By reducing the scale of the global crisis to the dimensions of a small rural town, and by personifying abstract social forces in the form of the town's inhabitants, the programme provides an accessible arena in which to address questions which do not find expression elsewhere. Frederick Jameson defined postmodernism some 30 years ago as an era in which culture is put to work as part of the economic system (1991). Culture no longer seeks to transcend the consciousness of the existing means of production but is an active player in maintaining and reproducing the patterns of economic activity and social relations. Of course, to some extent this is nothing new. The Victorian novel played an important role in educating the working classes into the proper (bourgeois) forms of behaviour. But under neoliberalism culture can no longer offer a perspective from outside our place within the means of production. There simply no longer is a temporal or geographical other to escape to.

Under liberalism, even as everything changes, everything remains the same. Jameson offers as an example of this paradox the typical English high street. The shop fronts may change continually, but the actual activity that is going on, and the economic, social and legal practices which define that activity remain unchanged. According to Jameson's understanding of postmodernity the past is henceforth only allowed to exist under particular terms, as nostalgia, decoration, to provide inert cues and symbols to be repackaged and reinterpreted in the service of the eternal now. When we want to put it away, we can, maybe to be forgotten forever, or maybe reused again at some future date, if it is useful. Can it sell a product? Can it legitimate current wars? Does it remind us how much more benign, caring and smarter we are now than we were in the past? Then bring it out. But there is no place for uninvited reminders of the past in this time beyond history. We have the power to decide when the past can be allowed to exist, what parts of the past and how long it will remain in the light. But the only history is that which, paradoxically, reaffirms the boundless now, which proves we are at the end of history, that change in the relations of productive forces is now impossible. There is not and never will be any escape, for there is nowhere to escape to.

In this sense, *The Returned* can be seen as a working through of a past that has come to haunt humanity, uninvited, and won't leave. The past in question is expressed in the crisis of climate change, which in embodying the emissions of prior as well as current industrial activity, has broken the illusion of the endless now which is so important to the success of liberalism. But we are not just living with the past of 150 years of industrial activity. We are living with the past embodied in the fossil fuels we are burning. It is as though the whole of life, all that ever lived, has turned up at the door, demanding to come in. This uninvited guest refuses to leave; yet the guest will not tell us what she wants. The full force of the reckoning is yet to be revealed. Like the angry and sullen parent, storing up some unknown punishment to be expressed at some unbidden time, we await in quiet terror for the inevitable storm.

That is what the town's people find so offensive about the return of the dead who, after all, are the children, partners, parents and friends of the living and so might expect to have been greeted with open arms and joy. They represent the past, and the past has no place in our present. The returned weren't invited, they won't leave and are not wanted. The returned struggle to accept that they are not wanted, that the past has no place there. But they remain, unforgivably and unforgivingly. Towards the end of the first season, a dam by the town starts to drain, and we see a previously submerged town starts to be revealed. The relevance of this drowned town to the existing one is not explained but one thing is clear. An unnatural act was committed so that the present-day town might live, flourish, grow and prosper. It is the consequences of those unnatural actions that have now taken on the form of the returned. It does not require too great a stretch of the imagination to interpret that unnatural act as industrial modernity, the consequences of which are the unravelling of natural systems as a result of the unprecedented changes these processes have wrought on the chemistry of the atmosphere.

By now, we all know the drill – rising global temperatures leading to melting ice sheets, acidifying oceans, droughts, storms heat waves and what have you. Except, when we look outside, it all seems normal. Yet, this normality is impregnated with a certain tension; whatever the weather today, it is difficult to escape the sense that all is not as it once was. The scene outside is imprinted with the past, the greenhouse gases that have been emitted building the rich lifestyles of the West. We know about explosions of methane blowing open huge craters in Siberia. Hundreds dying in India from unprecedented heat waves. Seemingly endless drought in California. And so, we wait, wondering when our time will come, how long until the next weather extreme. This new weather is history taking a form that can do anything but pose an existential threat to the current order far more powerful than any crisis of capital. Like the townsfolk of The Returned, we cannot escape anywhere, there is no planet B as the banners proclaim. And so, we try to carry on with our lives as normal, all the while the clouds, the wind, returning to us the whispered portent of a reckoning with our past.

4.5   Conclusion

The need to dehistoricise the climate issue is driven by class politics. If climate change were to become historical in the commonly accepted definition of the term, it would mean an all-encompassing change in the political economy of the world. To historicise climate change would be to explore the possibility of changing how the means of producing and reproducing our existence are changed irrevocably. Therefore, history must be understood to be at an end, and the injustices, hardships and suffering driven by climate change are to be addressed through technological change, rather than historical change. This situation exposes another contradiction at the heart of liberalism's response to climate change and reveals the working of liberal ideology; liberalism if nothing else, was meant to be a break from hidebound tradition and custom, and instead offered a chance for experimentation and creativity in imagining how we might live. That promise is now at an end.

Note

1 As the historian Michael Howard described it, the liberal conscience is 'not simply a belief or an attitude but also an inner compulsion to act upon it' (1978: 11, cited Dillon and Reid, 2009:1). Howard identifies liberal warfare as the key to this goal – liberal universalisation of war as a crusade with only one of two outcomes: endless war or the transformation of other societies and cultures into liberal societies and cultures (1978:128, cited Dillon and Reid, 2009: 3).

References

Adam, B., Beck, U. and van Loon, J. (2000). *The Risk Society and Beyond: Critical Issues in Social Theory.* London: SAGE.

Amy, D.J. (1984). 'Why policy analysis and ethics are incompatible', *Journal of Policy Analysis and Management*, 3(4), 573–591.

Anderson, P. (2017). *The H-Word. The Peripeteia of Hegemony*. London: Verso.

Beck, U. (1992). *Risk Society. Towards a New Modernity.* London: SAGE.

Bernstein, S. (2002). *The Compromise of Liberal Environmentalism*. New York: Columbia University Press.

Brown, K., O'Neill, S. and Fabricius, C. (2013). 'Interrogating transformation: Social science perspectives. In: World Social Science Report 2013'. Retrieved from https://www.oecd.org/fr/publications/world-social-science-report-2013-9789264203419-en.htm

Brulee, R., Carmichael, J. and Jenkins, J. (2012). 'Shifting public opinion on climate change: An empirical assessment of factors influencing concern over climate change in the U.S., 2002–2010'. *Climatic Change,* Vol. 114,169–188.

Carvalho, A., van Wessel, M. and Maeseele, P. (2017). 'Communication practices and political engagement with climate change: A research agenda'. *Environmental Communication,* Vol. 11, No. 1, 122–135.

Cook, D. (2018). *Adorno, Foucault and the Critique of the West*. London: Verso.

Dean, M. and Zamora, D. (2021). *The Last Man Takes Acid.* London: Verso.

Doulton, H. and Brown, K. (2009). 'Ten years to prevent catastrophe?: Discourses of climate change and international development in the UK press'. *Global Environmental Change*, Vol. 19, No. 2, 191–202.

Eagleton, T. (2016). *Culture*. Cambridge: Yale University Press.

Ellul, J. (1965). *The Technological Society.* London: Random House.

Feyerabend, P. (1978). *Science in a Free Society.* London: Verso.

Fischer, F. (1987). 'Policy expertise and the new class: A critique of the near conservative thesis', in Fischer, F. and Forester, J. (eds). *Confronting Values in Policy Analysis. The Politics of Criteria*. London: SAGE, pp. 94–127.

Foster, J.B. (2017). *The Long Ecological Revolution.* Retrieved from https://monthlyreview.org/2017/11/01/the-long-ecological-revolution/

Foucault, M. (2020). *Power: The Essential Works of Michel Foucault 1954–1984.* London: Penguin.

Ghosh, A. (2016). *The Great Derangement. Climate Change and the Unthinkable*. London: The University of Chicago Press.

Groys, B. (2012). *Antiphilosophy. An introduction.* London: Verso.

Habermas, J. (1998). *The Inclusion of the Other. Studies in Political Theory*. Cambridge, MA: MIT Press.

Hammond, P. (2018). *Climate Change and Post-Political Communication: Media, Emotion and Environmental Advocacy.* London: Routledge.

Holloway, J. (2019). *Change the World without Taking Power. The Meaning of Revolution Today.* London: Pluto Press.

Huitema, D. and Meijerink, S. (2010). 'Realizing water transitions. The role of policy entrepreneurs in water policy change'. *Ecology and Society*, Vol. 15, No. 2, 26.

Ingebritsen, C. (2002). 'Norm entrepreneurs: Scandinavia's role in world politics'. *Cooperation and Conflict*, Vol. 37, No. 1, 11–23.

Jameson, F. (1991). *Postmodernism, or The Cultural Logic of Late capitalism.* Durham: Duke University Press.

Kemp, L., Xu, C., Depledge, J. et al. (2022). 'Climate endgame: Exploring catastrophic climate change scenarios'. *PNAS,* Vol. 119, No. 34. doi: 10.1073/pnas.2108146119

Laski, H. (1936). *The Rise of European Liberalism.* London: Routledge.

Leiserowitz, A., Carman, J., Buttermore, N., Neyens, L., Rosenthal, S., Marlon, J., Schneider, J. and Mulcahy, K. (2022). *International Public Opinion on Climate Change, 2022.* New Haven, CT: Yale Program on Climate Change Communication and Data for Good at Meta.

Lukes, S. (2017). *Liberals and Cannibals: The Implications of Diversity.* London: Verso.

Malm, A. (2018). *The Progress of This Storm.* London: Verso.

Marx, K. and Engels, F. (1888). *The Manifesto of the Communist Party.* Moscow: Foreign Languages.

Moore, M.-L., Olsson, P., Nilsson, W., Rose, L. and Westley, F.R. (2018). 'Navigating emergence and system reflexivity as key transformative capacities: Experiences from a Global Fellowship program'. *Ecological Society*, Vol. 23, No. 2.

Noble, D. (2012). *Debating the End of History. The Marketplace, Utopia, and the Fragmentation of Intellectual Life.* London: University of Minneapolis Press.

O'Brien, K. and Sygna, L. (2013). 'Responding to climate change: The three spheres of transformation'. *Proceedings of Transformation in a Changing Climate*, 19–21 June 2013, Oslo, Norway. University of Oslo, pp. 16–23.

Orwell, G. (1943). *Can Socialists Be Happy?* Retrieved from https://www.orwellfoundation.com/the-orwell-foundation/orwell/essays-and-other-works/can-socialists-be-happy/

Orwell, G. (2001). *The Road to Wigan Pier.* London: Penguin.

Parenti, C. (2013). 'A radical approach to the climate crisis'. *Dissent Magazine.* Retrieved from https://www.dissentmagazine.org/article/a-radical-approach-to-the-climate-crisis

Peckham, M. (1995). *Romanticism and Ideology.* London: Wesleyan University Press.

Roberts, J. (2004). *Environmental Policy.* London: Routledge.

Rodney, W. (2018). *Walter Rodney's Russian Revolution: A View from the Third World.* London: Verso.

Ross, A. (1991). *Strange Weather: Culture, Science, Technology in the Age of Limits.* London: Verso.

Schmidt, J. (2001). *Disciplined Minds.* London: Rowman & Littlefield Publishers, Inc.

Sebald, W.G. (2014). *A Place in the Country.* London: Penguin.

Shove, E. (2010). 'Beyond the ABC: Climate change policy and theories of social change.' *Environment and Planning A*, Vol. 42, 1273–1285.

Stears, M. (2021). *Out of the Ordinary. How Everyday Life Inspired a Nation and How It Can Again.* Cambridge, MD: Harvard University Press.

Stoknes, P. (2015). *What We Think about When We Try Not to Think about Global Warming: Toward a New Psychology of Climate Action.* Vermont: Chelsea Green Publishing.

White, C. (2003). *The Middle Mind. Why Consumer Culture Is Turning us into the Living Dead*. London: Penguin.

Whyte, H. (1973). *Metahistory: The Historical Imagination in Nineteenth Century Europe*. Baltimore, MD: John Hopkins University Press.

Wodak, R. (2008). 'Discourse studies – important concepts and terms', in Wodak, R. and Krzyzanowski, M. (eds). *Qualitative Discourse Analysis in the Social Sciences*. Basingstoke: Palgrave Macmillan, pp. 1–24.

5 Guardrail 4: We have the technologies to solve climate change

Human progress is real, is expressed through technological change, and is an unassailable imperative of human existence. The alternative to this position is to plead that we go back to living in caves. We can solve climate change and keep the lights burning.

Liberal societies, ostensibly made up of citizens with agency and political freedom, are geared towards the promotion of scientific and technological agendas. These agendas are beyond the reach of democratic control. This chapter explores how these tension between freedom and technological determinism are manifested in climate change narratives. Who has ownership of the scientific and technological direction of travel? What social and political alternatives are displaced by looking to advances in science and technology for the answers to our problems? These are important considerations at this stage of the climate story. They are not considerations that feature as part of the climate discourse.

This chapter begins with an overview of the development and reproduction of the 'technological society' (Ellul, 1965). The second part of the chapter investigates implications for democracy of life in a highly technological society. The final part of the chapter examines how the language of climate change squares the circle of agentic individuals as the drivers of action on climate change in a world where humans are required to adjust to the demands of science and technology.

The term instrumental rationality refers to the idea that only that which can be quantified and measured is to be worthy of consideration. Max Weber contrasted instrumental rationality with value rationality. Value rationality is concerned with choosing the ends to which our efforts are directed, and for Weber was properly the domain of ethics, morality and politics. Instrumental rationality, on the other hand, is concerned only with the most efficient means of achieving a subjectively (e.g. privately) ordered list of ends or desires (see Kalberg, 1980, for an in-depth analysis of Weber's writings on rationality).[1] Considering climate change options within this typology, any attempt to resolve the problem by reconsidering the sort of world we want to live in (value rationality) is irrational, by dint of the fact that such considerations are beyond the calculus of instrumental rationality. The only (instrumentally) rational response is to find the most efficient means of reproducing the current social order. Such a constrained response to climate change only

DOI: 10.4324/9780429463488-6

appears reasonable because instrumental rationality has come to dominate over all other political considerations. The dominance of instrumental rationality destroys a society's ability to reason in what Adorno termed 'a negative dialectic' wherein our growing dependence on the technologies of instrumental rationality further erodes our ability to reason, which, in turn, deepens our dependency on technology (Wellmer, 1985). Reason is encouraged only in so far as it is needed for the operation of machinery (Ellul, 1965). In the end, all that remains is the logic of the machine and efficiency. Morality and ethics no longer exist as criteria for guiding human action. Whatever innovations are produced under prevailing patterns of power and privilege, come to be recognised uncritically as "progress" (Stirling, 2014).

This is a very deterministic lens through which to view our climate predicament. Such a strongly deterministic attitude is in marked contrast to the approach which sees society as made up of individuals with the agency and power needed to change the world, if only people would choose to use that power. The implication of technological determinism is that even were people to realise this latent power and agency, it would only ever be in the service of the machine. The material and philosophical manifestation of instrumental rationality's dominance goes under the name of techne, technique or the technological society. Techne is not simply a choice, one of many ways of understanding our world. It is the making of a world, a complete world (Castoriadis, cited Galarraga, 2015: 203). Ezrahi describes techne as a pre-theoretical horizoning (Ezrahi, 1994: 97). Any effort to shape the future will always and everywhere be defined by the demands and possibilities of technology. Technology is then an ideology – it is constitutive of the very possibility of experience (Ezrahi, 1994: 95). Heidegger goes further, to argue that although chronologically speaking, modern science precedes modern technology, metaphysically speaking technology is prior; 'technology beckons and shapes science, calls it into being in order to make nature "report" in a technological way, as a calculable coherence of forces' (Hediegger, 1954, cited Szerszynski, 2010: 19–20).

Ellul was concerned to understand the social adjustments needed for life in a world governed by technique, or what he termed the 'technological society' (Ellul, 1965). He used the concept of 'technique' to guide his investigation of how society has been affected by our internalisation of the machine's requirements. For Ellul, the shift to a society dominated by technique is the most fundamental transformation humanity has faced since the advent of agriculture (Boeckel, 2015: 235). Technique is an all-encompassing environment in which thought and action takes place – 'the totality of methods rationally arrived at and having absolute efficiency in every field of human activity' (Ellul, 1965: 190). We are trapped because only new technologies can solve the problems existing technologies are creating, as it is the determining factor of our society (Boeckel, 2015: 231). Technique is the process whereby the machine is integrated into society, in the process creating the kind of society the machine needs. Thus, in the technological society technique becomes the centre of society; civilisation is created for technique by technique and is exclusively technique. There is never any end to the improvements in efficiency; nor is there any end to the areas of existence which can be made more efficient. Technique is the promise of a limitless world. Our Year Zero is around 1780. It was then that the age of technological myths was born (Ellul, 1965) and the state began to accelerate the number of people brought under its control and coerced into producing a surplus for appropriation by elites (Scott, 2017: 152). It was around this time that we killed nature (Merchant, 1980), and liberalism was proffered as the mask for subservience to the machine, the story that sugars the pill of domination and destruction of the other, whether human or natural, leaving us all alone in the world. Since at least 1750, the idea of progress as a move towards a more just society was replaced by elites with the idea of progress as technological change (Ezrahi, Mendelsohn and Segal, 1994: 13). Technology as a substitute for politics didn't really take off though until the 1930s (Ross, 1991: 121). Herbert Hoover, an engineer, ran for president on an efficiency ticket, raising efficiency to the status of a moral category (ibid.: 123).

The promise of liberal technological societies is that the only limits anyone should have to face are the ones we wish to impose on ourselves (Lukacs, 2005). There shall be no ethical or moral judgement passed on the rule of technique (Boeckel, 2015: 233). The only moral imperative we need to observe is the drive to make all human activity more efficient, whether or not that activity is destructive. The idea of ecological limits or any other physical limits on human expression and identity has been anathema for the West since the second half of the 20th century. The post WWII period of relative social stability in the US and Europe saw the emergence of an "exclusionist" social paradigm, which excluded humans from the laws of nature (Porter and Brown, 1991: 2). The international agreements on limiting climate change are, by pushing the deadline out to 2050, an effort to sustain the limitless world by providing the space and time for a technological response to be developed. After all, if there was no dangerous limit, if all human forcing of the climate was considered unacceptably dangerous, then the risk causing activity would have to stop immediately.

5.1 Science against democracy

In arguing that the real political struggle is not between left and right, but between the political (the right) and the technocratic post-political, Zizek foresaw the contours of the climate debate in the West (Zizek, 2011: ix). Today, the only actors seeking to politicise climate change are on the right. The centre and centre-left wish to depoliticise climate change and sustain a consensus around the role of technology and science in delivering a net zero 2050. The goal is to decarbonise liberalism. The clear divide in climate politics today is not between climate deniers and everyone else, but between those who support the primacy of a science-led net zero by 2050 ambition, and those who seek alternative answers to the climate crisis. To express disquiet or question this science-led climate response is to be cast into the climate wilderness. Religion has, since the advent of science, become a choice in the West, whereas previously unbelief was not an option. Today, unbelief in science is no longer an option (Lahsen, 2007: 174).

Science and technology today are often presented as somehow separate from politics (Ezrahi, 2014: 30). Castree et al. reject this claim, arguing instead that science is always and everywhere political, and that to pretend otherwise is dangerous. They argue, this division supports the reproduction of "tornado politics". This is where

> crisis rhetoric ('we need to act now!') serves to suspend robust societal debate about future pathways. It leads researchers to focus only on the 'best' means necessary to reach given environmental goals in light of existing arrangements — thus leaving these arrangements relatively immune to questioning.
>
> (2014: 764)

Technology delegitimises alternative non-instrumental rationality and delegitimises genuinely political questions, because they are subjective and eternally contestable (Habermas, 1998: 106–107). Because science and technology are held to be non-political, they are not open to contestation, not subject to normative choices. They are ideological not only because they are not contested, but also because technology constrains the development and scope of belief to that aligned with the needs of technology. Because science and technology are not contested, this prevents consciousness of contrary evidence to the dominant position (ibid: 96). Though empirical evidence, cost-benefit analysis and quantification are intended to be a substitute for politics and opinion, evidence cannot fully supplant value judgements (Hammersley, 2013: 18). The objectivity promised by empirical evidence is, according to Passmore, offered up as a mask for a range of destructive and selfish values. Hence the insistence on focusing on what we can quantify, marginalising all other considerations (Passmore, 1978: 45). The truths that science discovers are themselves the product of political choices, because science can't happen without funding. Science becomes just another force of production, a moment in the circulation of capital (Jeffries, 2021: 101). Feyerabend argues the victory of science and instrumental reason was far from assured upon its first appearance (Feyerabend, 1978). In the 17th, 18th and 19th centuries, science competed with other views and ideologies and its methods and achievements were subject to an ongoing and widespread critical debate. However, whatever the initial barriers faced by the scientific mindset, its domain today is now complete and it faces no cultural resistance. The notion of science as a progressive force is not self-evident and the message requires constant repetition and defending. No holds are barred in the maintenance of this ideal of science because without continual technological innovation, economic progress would cease (Feyerabend, 1978: 82–83).

Science and technology are deterministic not only in the way they determine our activities day-to-day, but also in how they shape what future it is possible to imagine and in the inescapability of their continued dominance of the human condition. Ellul saw that by providing for all our needs, we will never be able to escape the technological society. Ezrahi et al. (1994) were moved to describe Ellul's "no exit" conclusion as the starkest of technological pessimisms.

Science and technology are the only forms of knowledge able to solve the problems they themselves have created. There is no longer any cultural resistance to the frame of technological inevitability. The path before us is clear – more science, more technology, giving us total understanding, total control. It is by now impossible to imagine putting boundaries around or limiting the role of science (Passmore, 1978: 1–3). There is no such thing as too much technology. The only dispute is who can speak the language of technique most proficiently – industry, politicians, campaigners? (Reinecke, 1984; Ross, 1991: 98).

Though liberalism is defined as a separation of powers between the state and other institutions, in fact, science and the state are conjoined (Feyerabend, 1978: 74). This results in a climate discourse which seeks to centralise decision-making under the state, and prioritises a scientific response because the science can be directed by the state, as the primary funder of scientific research. Scientific ideas are not subject to a public vote. Consequently, alternative non-technological and non-scientific responses to climate change are side-lined. For example, relocalisation is not on the climate policy agenda, even though (or more precisely, because) it is less technological, less competitive and more egalitarian than existing social relations (Fischer, 2017: 226). An impulse for local governance has been a feature of all revolutions (ibid: 233) but does not feature as an aspect of the 'fundamental change' seen as necessary for living sustainably. Technological change is the only answer offered to the problems it has itself generated (Ross, 1991: 98). There is no cultural resistance to the frame of technological inevitability, nor can there ever be (ibid: 119). Meanwhile we continue waiting passively for science to deliver on its promise of slaying the climate monster so we can return to a place of ahistorical existence, the stability of a universal liberal rationality, a bourgeois utopia where technology rules, politics is dead and choice is exercised through the market. Any problems that arise are particular, idiographic, and will be dealt with by the relevant area of expertise.

Whilst our technological society provides the state with immense power, it disempowers citizens, and dismisses non-scientific responses to environmental destruction. Such involvement as is encouraged is often as citizen scientists, which whilst offering the promise of exploring democratic processes for building a meaningful response to climate change, in fact, packages the environment as a science issue, which will be encountered by "citizens" as a "science" issue (Phillips, Carvalho and Doyle, 2012). The World's Fair Futarama building of 1939 showed America in 1960 and visitors to the display were given "I have seen the future" badges, so at this stage people began to be moulded into passive observers of a future planned by others (Ross, 1991: 131). Technology was sold as the instrument of liberation for the proletariat. It needed only technological progress for the proletariat to free itself a little more from its chains (Ellul, 1965: 73). Today, the future has become the natural habitat of (and has been colonised by) technocratic elites (ibid.: 172). This future has a predetermined technocratic outcome, to which the facts must be made to fit. We don't control technological progress, no one controls technological progress, it is a force of nature (Dickens, 1992).

5.2   Selling technological responses to climate change

The rise of a scientific and technological society has met with continued disquiet from a public that no longer feels at home in the world science has made. Twenty years ago, the European Commission warned:

> Our society is faced with the challenge of finding its proper place in a world shaken by economic and political turbulence ... science, technology and innovation are indispensable to meet this challenge. However, there are indications that [their] immense potential is out of step with ordinary citizens.
>
> (European Science Communication Network, 2001)

In response, the Commission proposed 38 actions, involving researchers, media professionals and the Commission itself, that were vital if the public were to have confidence in and support their scientists in providing solutions to the problems facing the EU. The problem was not with the reliance on science and technology, but how it was being sold. As Passmore would have it, the default position is that the scientist is pure, but her work is seized upon by the impure (1978: 27). Unfortunately, it appears that whilst the rule of technique has created the ecological damage we are now witnessing, it has yet to offer a solution. In reviewing our prospects, Smil is keen to demonstrate the actual material scale of the challenge we face. He records that

> the impact of technological advances on fossil fuel remains as much as it ever was. Over the first 20 years of the 21st century, the amount of energy derived from fossil fuels declined from 87% to 85%, and most of that decrease came from expansion of a relatively old technology, hydroelectric power.
>
> (Smil, 2022: 216)

The failure to act on climate change is attributed by some observers to a lack of public and political will to make full use of our scientific and technological power. 'Humanity already possesses the fundamental scientific, technical, and industrial know-how to solve the carbon and climate problem for the next half-century' (Pacala and Socolow, 2004: 968). Promising science can "solve" climate change requires some rather unscientific approaches to representations of climate risk. White makes the point that science's fantasy is that it can put a scientific boundary on destruction (White, 2007: 44). Pacala and Socolow equate "solving" climate change with limiting atmospheric concentrations of carbon dioxide to between 450 parts per million (ppm) and 550 ppm (2004: 968). About 450 ppm is expected to lead to warming in excess of 2°C, and has been described as a 'prescription for disaster' (Hansen and Sato, 2011). So, on the basis of this calculation, it would appear impossible to avoid disaster in a techno-industrial complex (Bahro, cited Fischer, 2017: 244). But rather than look outside of science and technology for an answer to climate change, we double down on the effort to make climate change governable through market and technological solutions (Oels, 2005). The more our plight worsens, the more desperately we turn to machines for salvation (Hine and Kingsnorth, 2015).

This deepening dependence on technology to solve environmental problems is celebrated by advocates of ecological modernisation. Ecological modernisation commands the climate policy landscape, provides the framework for coalitions between civil society and political and corporate actors (Schaefer et al., 2015: 250). Though there are – in theory at least – a wide range of possible perspectives on climate change, nearly all the organisations gaining political and media access are ones espousing ecological modernisation (ibid.: 252). There is a certain irony in the liberal championing of ecological modernisation. A technological capitalist civili-sation requires the global homogenisation of people's desires and socio-economic relations, in order to create a global order that can effectively align production and consumption at a massive scale (Anderson, 2017: 40). That is why liberalism, argues Foster (2017), when confronted by climate change, takes on the mantle of ecomodernism, which shares with modernism a denial of natural limits to capital accumulation. This homogenisation does not bother the champions of individual-ism and freedom. This is because liberalism cherishes diversity not per se, but liberal diversity, 'that is, diversity confined within the narrow limits of the liberal model of human excellence' (Parekh, 1994, cited Lukes, 2017: 13).

The key note report on sustainable development from 30 years ago laid out the ecological modernisation blueprint when it promised, 'technology and social organization can both be managed and improved to make way for a new era of economic growth' (Brundtland, 1987: 1). Ellul thought that technology and tech-nique are considered sacred in modern Western societies. Technology is the God which saves, though we don't know how it will save us, but we must have faith (1965). His words presaged by more than 50 years the sentiments expressed by Sweden's Deputy Prime Minister, Isabella Lovin in an article stressing the need for ambitious action to reach net zero targets. Lovin wrote 'Here, we must trust to innovation, the speed of which continues to amaze' (Lovin, 2018). So, innovation (God) moves in mysterious ways but we must not question its ability to provide and keep us safe, and must simply have faith. The path before us is clear – more science, more technology, giving us total understanding, total control (Passmore, 1978: 1). Though the claims of technological salvation from climate change are repeated with ever greater vigour, they are made in the absence of any evidence base. They have become ideological, and dangerous. Ideologies begin to deterio-rate and become dogmatic once they quash the opposition. Their victory is their own downfall (Feyerabend, 1978: 81). The victory of technique may also be humanity's downfall. Any stabilisation new technologies can offer (and they have offered none to date as regards climate change) can only ever be provisional whilst the forces creating the emergency are in conflict with the logic of natural systems (Streeck, 2014: xv).

In terms of climate change, the dependence on technology, combined with the uncertain and complex nature of global environmental problems, helps ensure the importance of technical expertise in influencing and shaping international policies. Bernstein argues that the role of technical and scientific knowledge in environmental policy formation has been greatly exaggerated. Instead, existing social structures are a far more powerful influence on policy decisions than science and technical expertise (2002: 124). Existing social structures and associated norms guide the consensus reaching process and appropriate behaviour amongst the small scientific epistemic community that is politically empowered and in a close relationship with policy-making institutions. Bernstein argues the ideas behind liberal environmentalism did not originate among scientists but did increasingly influence scientific work (2002: 125–127). While scientists did play a role in bringing problems to the attention of politicians, they played a remarkably minor role in the development of the policy responses (ibid). After 1987, governments appeared to make a more concerted effort to rein science in rather than allow freewheeling scientists to dictate the environmental agenda (Bernstein 2002: 156). Today, the mainstream of climate change debate is both apolitical and ahistorical. Transformational change simply means electrifying the energy system. There is only "now", a moment on a linear path to a technologically better future, within a political stasis. That political stasis, a requirement of the economic status quo, has profoundly dangerous implications for our ability to act on climate change. Nietzsche calls the inability to change, when life calls for transformation, decadence (cited Anfinson, 2107: 225). This decadence characterises intellectual life in the face of climate change, an academia rendered moribund by a frightened conventionality, science reduced to mere pedantry (Caudwell, 2018: 78). And yet, all efforts on climate change are first and last directed at maintaining and extending this way of life. Currently, the only option seems to be the eternal finessing and parsing of measurements, climate sensitivities and probability distribution functions. This is being written in the summer of 2022, in the middle of a global heatwave and Europe wide drought; in the UK, this week is forecast to be the hottest ever on record in the UK, and warnings have been issued for people to stay indoors with their curtains drawn between 11AM and 3.00PM. During the (at that time) record breaking UK heatwave and drought of 2018, the leading liberal paper in the UK carried a story which cited scientists claiming we should be careful about overstating the role of climate change in these extreme events (McKie, 2018). To imagine the most pressing task facing humanity is to spend more time and money finessing the fractions on what proportion of the cataclysm can be attributed to climate change is profoundly decadent. One is unable to avoid the question, 'Why get stuck in such perplexities at all?' (Winner, 1986: 175). We know all weather is now being experienced in a climate changed by human activity. Let's just call it, and work out what to do. However, the fetishisation of science, numbers and targets diverts attention away from questions about the political and social order (Smith and Elliot, 2007: 2). Scientists and activists are willing to go along with this farrago in order to keep a seat at the table of power (Newell, 2000). It is from this position that the concept of a post-normal science has emerged, which imagines – despite all the evidence to the contrary – that science can be made to sit alongside a range of other considerations as part of a democratic decision-making process with an extended peer review network (Funtowicz and Ravetz, 1993).

5.3 Conclusion

Technology offers the promise of progress, but traps us in a cycle of dependency. We need more technology to solve the problems created by existing technologies. The initial promise of emancipation offered by technology has now receded. Technology distances us from direct experience of the world and each other and so disempowers us. Technological change and science are not under democratic control. Humanity is reduced to being passive bystanders of whatever changes technology brings upon us and the world. Discussions of climate policy under liberalism have never, and could never, countenance a move away from our reliance on new technologies as a way of limiting climate change. Instead, what is needed are more new technologies. In the place of our old Gods, we now have technology – distant, mysterious, bringer of life and death. Technology tightens its grip on our imagination day by day because we live in a world ruled by technology and technique – the search for ever more efficient ways of reaching goals defined by instrumental rationality.

Note

1 It is worth noting here that day-to-day behaviours may also be shaped by another dimension of rationality – bounded rationality. This is the rationality that emerges in the face of constraints such as limited information, and limited time (Simon, 1982).

References

Anderson, P. (2017). *The H-Word. The Peripeteia of Hegemony*. London: Verso.

Bernstein, S. (2002). *The Compromise of Liberal Environmentalism*. New York: Columbia University Press.

Boeckel, J.V. (2015). 'Never mind where, as long as it's fast', in Kingsnorth, P. and Dougald, H. (eds). *Techne. www.darkmountain.net*. Croydon: CPI Group, pp. 229–243.

Brundtland, G.H. (1987). *Our Common Future: Report of the World Commission on Environment and Development*. Geneva, UN-Dokument A/42/427.

Castree, N., Adams, W., Barry, J. et al. (2014). 'Changing the intellectual climate'. *Nature Climate Change*, Vol. 4, 763–768.

Caudwell, C. (2018). *Culture as Politics. Selected Writings of Christopher Caudwell*. Margolies, D. (ed). London: Pluto Press.

Dickens P. (1992). *Society and Nature – Towards a Green Social Theory*. New Jersey: Prentice Hall.

Ellul, J. (1965). *The Technological Society*. London: Vintage.

European Science Communication Network. (2001). *Science and Society Action Plan*. Retrieved from https://esconet.wordpress.com/about/

Ezrahi, Y. (1994). Technology and the illusion of the escape from politics', in Ezrahi, Y., Mendelsohn, E. and Segal, H. (eds). *Technology, Pessimism, and Postmodernism*. Sociology of the Sciences, Vol. 17. Dordrecht: Springer, pp. 29–37.

Feyerabend, P. (1978). *Science in a Free Society*. London: Verso.

Fischer, F. (2017). *Climate Crisis and the Democratic Prospect: Participatory Governance in Sustainable Communities*. Oxford: Oxford University Press.

Foster, J.B. (2017). 'The long ecological revolution'. *The Monthly Review*. Retrieved from https://monthlyreview.org/2017/11/01/the-long-ecological-revolution/

Funtowicz, S. and Ravetz, J. (1993). 'Science for the post-normal age'. *Future*, Vol. 257, No. 7, 735–755.

Galarraga, M. (2015). 'The making of a world. Climate engineering as creation'. *The Dark Mountain Project*, No. 8. Autumn 2015, 202–210.

Habermas, J. (1998). *The Inclusion of the Other. Studies in Political Theory*. Cambridge, MA: MIT Press.

Hammersley, M. (2013). *The Myth of Research-Based Policy & Practice*. London: SAGE Publications Ltd.

Hansen, J. and Sato, M. (2011). 'Paleoclimate implications for human-made climate change'. *NASA Goddard Institute for Space Studies and Columbia University Earth Institute*, New York. Retrieved from http://www.columbia.edu/~jeh1/mailings/2011/20110118_MilankovicPaper.pdf

Hine, D. and Kingsnorth, P. (2015). 'The draw knife and the drone'. *Dark Mountain Project*. Croydon: CPI Group. Retrieved from www.darkmountain.net.

Jeffries, S. (2021). *Everything, Everywhere, All the Time*. London: Verso.

Kalberg, S. (1980). 'Max Weber's types of rationality. Cornerstones for the analysis of rationalization processes in history'. *The American Journal of Sociology*, Vol. 5, No. 5, 1145–1179.

Lahsen, M. (2007). 'Trust through participation? Problems of knowledge in climate decision-making', in Pettenger, M. (ed). *The Social Construction of Climate Change. Power, Knowledge, Norms, Discourses*. London: Routledge, pp. 173–196.

Lovin, I. (2018). To lead on climate, countries must commit to zero emissions'. *The Guardian*. Retrieved from https://www.theguardian.com/environment/2018/apr/17/to-lead-on-climate-countries-must-commit-to-zero-emissions

Lukacs, J. (2005). *Democracy and Populism: Fear and Hatred*. New Haven, CT: Yale University Press.

Lukes, S. (2017). *Liberals and Cannibals: The Implications of Diversity*. London: Verso.

McKie, R. (2018). *The Big Heatwave: From Algeria to the Arctic. But What's the Cause?* The Observer, 22nd July 2018. Retrieved from https://www.theguardian.com/world/2018/jul/22/heatwave-northen-hemisphere-uk-algeria-canada-sweden-whats-the-cause

Merchant, C. (1980). *The Death of Nature. Women, Ecology and the Scientific Revolution*. New York: Harper and Row.

Newell, P. (2000). *Climate for Change: Non-State Actors and the Global Politics of the Greenhouse*. Cambridge, MA: Cambridge University Press.

Oels, A. (2005). 'Rendering climate change governable: From biopower to advanced liberal government?' *Journal of Environmental Policy & Planning*, Vol. 7, No. 3, 185–207.

Pacala, S. and Socolow, R. (2004). 'Stabilization wedges: Solving the climate problem for the next 50 years with current technologies'. *Science*, Vol. 305, No. 5686, 968–972.

Passmore, J. (1978). *Science and Its Critics*. London: Routledge.

Phillips, L., Carvalho, A. and Doyle, J. (2012). *Citizen Voices: Performing Public Participation in Science and Environment Communication*. Bristol: Intellect Books.

Porter, G. and Brown, J.W. (1991). *Global Environmental Politics*. Boulder, CO: Westview Press.

Reinecke, I. (1984). *Electronic Illusions. A Skeptic's View of Our High-Tech Future*. London: Penguin.

Ross, A. (1991). *Strange Weather: Culture, Science, Technology in the Age of Limits*. London: Verso.

Schaefer, B., Brulle, R. and Szasz, A. (2015). 'Civil society, social movements and climate change', in Dunlap, R. and Brulle, R. (eds). *Climate Change and Society*. Oxford: Oxford University Press, pp. 235–268.

Scott, J. (2017). *Against The Grain. A Deep history of the Earliest States.* Yale, NH: Yale University Press.

Smil, V. (2022). *How the World Really Works*. Dublin: Penguin Random House.

Smith, D. and Elliot, P. (2007). 'Hazardous waste and technological risk: The limits of science in decision-making'. *European Environment*, Vol. 2, No. 1, 1–4.

Stirling, A. (2014). 'Transforming power: Social science and the politics of energy choices'. *Energy Research & Social Science.* Vol. 1, 83–95.

Streeck, W. (2014). *Buying Time: The Delayed Crisis of Democratic Capitalism*. London: Verso.

Szerszynski, B. (2010). Reading and writing the weather. *Theory, Culture & Society*, Vol. 27, Nos. 2–3, 9–30.

Wellmer, A. (1985). 'On the dialectic of modernity and postmodernity'. *Praxis International*, Vol. 4, 4.

White, C. (2007). *The Barbaric Heart. Faith, Money and the Crisis of Nature*. California: PoliPoint Press.

Winner, L. (1986). *The Whale and the Reactor. A Search for Limits in an Age of High Technology*. Chicago, IL: University of Chicago Press.

Zizek, S. (2011). *Living in the End Times*. London: Verso.

6 Guardrail 5: New stories will save us

We need to find new stories that can help humanity live sustainably. These stories show how the whole of humanity can align with the norms of western liberal democracies, and live clean, smart and prosperous lifestyles without dangerous emissions of greenhouse gases.

How far can language take us in the changes we need to transform the future? Compelling stories of the future are important rallying cries for social change movements. But such stories only successfully inform social change movements when the goals of those movements are aligned with the core imperatives of the state. Any time a social movement's goals come into conflict with the core imperatives of the state, the movement loses (Dryzek, Downes and Hunold, 2003: 2). These core imperatives are domestic order, external competition, revenue, economic growth and legitimation, and offer only limited opportunity for democratic intervention (ibid: 3). Unfortunately, these are exactly the areas of state activity that need to be confronted if the IPCC's calls for fundamental change are to be heeded. Dryzek et al. argue that the environmental movement has learnt to limit their radicalism to not impinge on these core imperatives, and has no desire to take control of the state, either peacefully or by any other means. As a result, the authors conclude there are no green states (ibid.: 11).

This chapter examines the extent to which stories can save us through non-conflictual change. Through a close examination of two non-academic books rooted in the social sciences, (Stoknes, P. (2015). *What We Think about When We Try Not to Think about Global Warming: Toward a New Psychology of Climate Action* and Evans, A. (2016). *Eden 2.0: Climate Change and the Search for a 21st Century Myth*), this chapter also explores what kinds of stories are imagined to offer the promise of salvation from the climate crisis. It is argued that liberal climate stories are premised on the belief that sufficient people will experience sufficient change in their beliefs to help bring about a smooth and conflict free transition to a net zero capitalist economy.

Despite repeated warnings from scientists, and much encouragement, sustained and substantive changes in the behaviours of individual citizens have not been forthcoming. Rather than look to corporations, the political system or neoliberal economics for change, the social sciences seek to understand how to generate a

DOI: 10.4324/9780429463488-7

measurable increase in reported levels of public concern about climate change, or the adoption of energy-saving behaviours. A considerable research effort has been dedicated to documenting the effects of different linguistic choices on public engagement with energy and climate change. This burgeoning industry of psychological research has spawned many different insights into how best to communicate climate change. Yet, the debate about what frames to use, or indeed whether framing has any impact at all on how people think and feel about climate change, remains a live one. Bernauer and McGrath (2016) were not able to find any robust empirical evidence that alternative framings of climate policy increase public support for greenhouse gas mitigation. A later meta-review of studies into framing concluded the opposite, suggesting framing does impact people's intentions to change their behaviours, and support for climate policies (Li and Yi-Fan Su, 2018). The confusion arises from the different frames that are tested, the different audiences they are tested with, the unnatural context in which people encounter the frames, and the way the results are interpreted. Studies may overestimate framing effects on attitude change, since research conditions do not correspond to how most members of the public encounter information about climate change in the real world. By way of the news, social media, or conversations, individuals are likely to encounter multiple, often conflicting or competing frames (McCright et al., 2016; Nisbet et al., 2013). One thing is clear amongst all this confusion – people have not fundamentally transformed their lives, behaviours or values.

Alongside the search for the correct frame to communicate climate change, is the idea that some metaphors and analogies might be more effective than others. As I described in an earlier work (Shaw, 2016), metaphors and analogies are important to thinking and acting in the world and have an important role to play in any narrative. They can enable as well as constrain the ways we think about policy issues, especially with regard to largely abstract, complex and seemingly intractable problems like climate change. Metaphors have an especially important role to play in anchoring novel phenomena in familiar and shared ideas (and hence language and culture). As was discussed in the Introduction, anchoring describes the means by which people come to understand an unfamiliar event. People can only make sense of the world by finding ways to reconcile their beliefs with some set of facts about how reality must operate. To anchor an object is to fit it into an existing system of classifications, is to name it and relate it to other objects in the system. Hence anchors allow groups to make sense of novel risks by classifying and naming the threat, making the unfamiliar familiar (Washer and Joffe, 2006: 2143). Metaphors, in providing an alternative framing for novel and abstract phenomena, are powerful anchoring devices (Wells, 1987: 443). It has been claimed that all understanding comes as the result of the perception of one thing through another. In other words, You don't see something until you have the right metaphor to let you perceive it. As with any other form of language, the choice of metaphors is often a political act. Metaphors have not only perceptual and cognitive, but also performative and political force, as they commit those who create and use them to accepting a system of standard beliefs or commonplaces associated with them. Consequently, metaphors have an important role in providing a shared and stable set of concepts across a range of centralised institutions with transglobal reach. These international organisations often set and implement key rules within national governments; create, channel and disseminate knowledge; shape dominant discourses; frame problems and solutions; influence negotiations through their ideas and expertise; and oversee the implementation of projects on the ground.

Though language has an important role to play in justifying and giving meaning to our social activities, this is largely a post-hoc relationship. What we do as a society is the right thing to do because the stories we tell ourselves as a culture legitimate the actions already taken, or already committed to. Climate change stories become not the fuel for political struggle, but a way of maintaining the status quo, and a substitute for political change. Stories of our net zero future are not meant to inflame rage, incite unrest. They are meant to reassure, control, soothe and pacify. They are stories, whose purpose is to displace the appetite for the fight needed to safeguard our survival (Anfinson, 2018: 210). Consciousness is a product of material conditions. Stories, on their own, are unlikely to bring about the fundamental political change demanded by ecological and climatic collapse.

6.1   Can new stories create new worlds?

Language is the instrument of our mediation between consciousness and the world we inhabit (Whyte, 1973: 88), making our relationship with the world primarily a relationship with a constructed world, rather than a world in which natural process flow through us unshaped by language. The invisibility of the interests and power relationships inherent in linguistic performances (Roberts, 2004) makes language a powerful device for building consent to policy decisions (Newell, 2000: 77). The construction of reality through language (Willig, 1998) means we can describe climate change as a quasi-reality (Bray and Shackley, 2004: 2). It therefore seems possible to imagine that stories are part of the changes needed for building a new world. Where stories appear in that process – as driving forward the change or simply as a way of making sense of change once that change has happened – is less clear. Marxism stresses that whilst changes in the economic base (relations of production) can bring about change in the superstructure (consciousness), this is a one-way process. Efforts to bring about changes in consciousness in order to bring about changes in the economic base will fail. So, language cannot in and of itself – contrary to some of the assumptions underpinning the 'narrative turn' in social sciences (Goodson and Gill, 2011) – effect the social and economic changes required to limit anthropogenic climate change. Jackson (1994) argues that the social sciences have, in their obsession with language, managed to convince themselves that economic realities and class divisions can be wished away simply by finding language that pretends they don't exist. There is undoubtedly an air of mystery and magic in accounts which argue that changing the stories is the first step to creating the new world. Why those with power will roll over and peacefully acquiesce to a world in which their control and privilege is taken away from them is never explained.

6.2 Culture as control

Schmidt (2008) identifies two forms of discourse in an institutional context: coordinative and communicative. While coordinative discourse is used among policymakers, communicative discourse 'consists of the individuals and groups involved in the presentation, deliberation, and legitimation of political ideas to the general public' (Schmidt 2008: 310). The two books discussed in this chapter are communicative discourses in the sense described by Schmidt. Both books promote the claim that climate change can be solved without revolution, and within the confines of the liberal worldview. The books are not books that deal with climate change science, but instead are from the social sciences, and make claims about how a change in the use of language will help solve climate change. The books have been chosen for analysis because the authors are socially and historically situated actors who seek to universalise their own values through climate change. In this case, those values are those of the liberal bourgeois.

One objective of the communicative discourses identified by Schmidt is containment, consisting of boundary setting and tailored framing (Schmidt 2008: 310). Boundary setting concerns the delimitation of the possible. Due to the political circumstances (an open and free democracy), disregarding alternatives is not an option. Therefore, the speaker sets the boundaries by acknowledging the possibility of applying alternatives (i.e. clearly distinguishable choices) but only to delegitimise them (Mair 2019). These discourses have substantial effects on the world because they justify certain policy responses, define the truths upon which one must base political action (Burr, 1998) and hence, in respect of climate change, dictate what harms will be experienced by who. These discourses do not on their own bring new worlds into existence, but instead prevent the emergence of new futures by justifying the existing conditions, and denying the possibility of alternatives. This is a relatively easy task, for it is human nature to imagine that 'as we and our surroundings are so must it have been always and everywhere' (Schlegel, cited Peckham, 1995: 56).

Culture controls behaviour and has as its principal task the maintenance of behavioural stability by circumventing the use of force, via seduction and intimidation (Peckham, 1995: xvi). The exercise of power is best hidden, if we are to continue to consider ourselves free, and it is for this reason that power works by fiction, ceremony and charade (ibid.). For Peckham, ideology is the conscious application of culture to maintain power (1995: 53). Revolution is the very opposite of this. It is a ripping apart of the curtains, a brutal revealing of the truth of power (ibid.). The Russian revolution was a war on signs, a process of removing all reference to Tsarist rule (ibid.: 59). Today, we live in a world that removes all signs of an alternative to liberal responses to climate change. The culture industry promotes conformity to prevailing standards and norms, undermining familial forms of socialisation (Cook, 2018: 41). Marketing and advertising are deeply ideological, and appeals to popular notions of common sense routes to happiness, satisfaction and a good life (ibid.: 152). It also deadens energies and the capacity to imagine any other world (ibid.). In addition, 'consumer society (objects, products, advertising) for the first time in history, offers the individual the opportunity for total fulfilment and liberation' (Baudrillard, 1968, cited Poster, 1998: 12).

Language (for the purposes of this chapter, in the form of books) as cultural objects, can control behaviour by instilling the values of the bourgeois liberal into the hearts and minds of the public. Culture creates an ethical world (White, 2003: 42), attributes positive values to behaviours and attitudes that are aligned with the interests of bourgeois liberalism. Spark, discussing reaction to Moby Dick, talks of an upper class project in the humanities that took aim at the unpredictable political imaginations of the newly emancipated lower orders (2002: 7). Spark goes on to record that the diffusion of literacy made possible through printing had been an object of anxiety for ruling elites ever since the Reformation (ibid: 19). It is this same principle which explains the emergence of the idea that humanity needs 'new stories' or 'new myths for the 21st century' to guide our response to climate change. It is important that they are the right stories.

For these right stories to be interpreted as the right stories by the "lower orders", societal actors need to act as "norm entrepreneurs," mobilizing support for particular ways of talking about and responding to social issues (Ingebritsen, 2002: 12). The field of critical discourse analysis (CDA) assumes that creating consent for a particular configuration of political power stems from the ability to articulate and set the terms of a discourse (Bäckstrand and Lövbrand, 2019: 125), delimiting what it is possible to speak at a given moment (Ramanzanglo 1993; 197). These norms are reproduced on a day-to-day basis by people articulating and defending those norms (even if they don't believe in them) in order to achieve benefits and avoid costs (Cass, 2007: 25). Understanding how discourse is used to facilitate the exercise of power, CDA scholars argue, is the first step to building new discourses and through these new discourses, change the distribution and exercise of power.

The majority of activists, professional communicators and others who generate and distribute stories of our climate futures are using language in ways that will avoid a hostile reception from powerful actors. They know what is acceptable, what is seen as reasonable and what is considered beyond the pale. In his study of political responses to air pollution, Crenson explains how power relations act to exclude certain options and responses from the frames employed by institutions (Crenson, 1971). The fingerprint of these 'non active forms of power' have been identified at play within the language used by environmental campaigners. It is these non-active forms of power that explain why despite survey after survey showing the public want corporations and governments to lead, nothing ever changes. Like a comatose patient lying in its hospital bed, the public are desperately trying to communicate to the doctors what they need, but the doctors either don't understand or don't care what the patient is trying to tell them.

6.3   Creating orderly transitions through stories

When they write 'the roots of our crises lie in the stories we have been telling ourselves', Kingsnorth and Hine overstate the power of stories to create political change (Kingsnorth and Hine, 2009). They are closer to the truth in their argument that our current stories are made possible by, and intended to hold in place, the very thin veneer of civilisation, underneath which lies a churning maelstrom of chaos and primal desire. This desire to maintain social order by diverting our dark desires into the relatively safe space of free markets is an issue that has been interrogated by Amitav Ghosh, who in 2016 asked why it was "serious" literature never included climate change? The importance of climate change to culture cannot be overstated, claims Ghosh. Culture generates desire for the products and actions driving climate change (2016: 9–10), and currently the absence of climate change from literature suggests a culture that is actively concealing this most urgent of all issues facing humanity (2016: 11). And the reason that climate change is concealed in literature is because this most bourgeois of art forms – the novel – is concerned with narratives that describe a world aligned with the calm, predictable regularity so treasured by the bourgeoisie (Ghosh, 2016: 19). Disruptive climate change impacts, and disruptive climate change policy, have no place in such a world.

Marx, writing about literary men, noted that what makes them representative of the petty bourgeoisie is the fact, in their minds they cannot transcend limits which shopkeepers cannot transcend in life, that they are therefore driven theoretically the same tasks and the same solutions to which material interest and social position drive the latter in practice (Prawer, 2011: 182). Hence stories are a means by which climate change discourse is depoliticised, by removing any reference to alternative forms of political organisation. They are also a means of imposing order on the world, show that human affairs are best managed through higher rational thought and planning (Sebald, 2014). Culture, in the form of stories, is supported in its quest for order by science. Scientific knowledge imposes order on disorder, reduces non-identity to identity, and is motivated to create a stable order in its own image by a fear of the chaotic and unclassifiable (Dews, cited Cook, 2018: 26). Progressives such as Milton, Locke, Diderot, Adam Smith, Thomas Paine, Condorcet and Thomas Jefferson, believed that through government regulation of markets and industry, and a partial redistribution of wealth and resources and incorporation of the dissonant elements (labour unions and minority groups), it would be possible to reinstate the equilibria thought to be lost during the transition from a peaceful and integrated agrarian society to the urban and industrial nightmare of market-induced greed, exploitation, economic strife, and anomie (Spark, 2001: 11). Spark is interested in the challenge the book Moby Dick posed to this order through the figure of Captain Ahab, 'whose search for the truth deviates from the beaten path and follows the facts wherever they lead, without regard for order and stability' (2001: 36).

We are all susceptible to stories because they separate the messy world into simple binaries of good and evil, right and wrong (ibid.). It is with this lens that we analyse two recent books that lean heavily on the idea of myth to articulate a vision of humanity's climate change free future. The first book – written by an ex-civil servant in the UK government – argues that we need stories and myths to make sense of the climate science, to turn those facts into an agenda for action that engages the public (Evans, 2016). The second book, written by a psychologist and economist, addresses the same theme; using stories as vehicles for facts which 'create positive solutions' (Stoknes, 2015). Both books speak to the liberal principles of promising an answer to climate change that does not require social conflict, and rests on the assumption that the solution will ultimately be technological. These two books are obviously not the only books that lay out thoughts on how to respond to climate change. They do not define the field, and they were chosen for analysis as one of the first steps I took in writing this book (which was over 4 years ago, hence them both being some 5–6 years old at the time of this book's publication). The main appeal of these books for this analysis is that the two white male authors are so explicitly exemplars of the bourgeois liberal ideal type that has commandeered the climate debate to date, whilst professing to be in the business of social transformation.[1]

6.4   Eden 2.0: Climate change and the search for a 21st century myth

'Our most respected leaders have also been able to communicate at the level of myth in the sheer grandeur of the stories they tell and in what they are able to summon forth from their listeners as a result. My great grandfather, Charles Wilson, was Winston Churchill's doctor throughout the second world war, and was in awe of Churchill's extraordinary power of storytelling.'

Evans obviously has deep roots in upper-middle class bourgeoise life. That he is speaking from a particular class perspective is nowhere admitted, or recognised as an issue relevant to the claims made. That he draws on the deified symbolism of the figure of Winston Churchill to show the importance of strong leadership demonstrates the book is rooted in the same structures, values and worldviews that have dominated national discourse over the time that climate change has developed into a crisis. That national discourse has reified Winston Churchill into an idol. The ideas Winston Churchill articulated and represented are obviously of a piece with the world that will 'solve' climate change.

The author's milieu is, just like his grandfather, in serving the needs of a Western imperial order.

'For almost for as long as I can remember, I wanted to work on global environmental issues. And once I'd seen my first episode of *The West Wing*, I knew I wanted to do it as a political adviser inside government.'

The framing of climate change as a "global environmental" issue communicates two ideas about climate change which align it with the principle of controlling people through the depoliticising of the issue. The first of these is the claim that climate change is a global issue. No one would argue that the climate is a global phenomenon and changes will be felt across the globe. But, in wanting to "work" on it as a global issue is to argue that the author wishes to universalise bourgeois liberal values through action on climate change. That the inspiration for this move came from The West Wing, the 'liberal's television bible' (Savage, 2017) is perhaps to be anticipated.

> 'In 2003, I became one of two special advisers to the British Secretary of State for International Development (first Valerie Amos, and then Hilary Benn) in Tony Blair's government.'

The author's willingness to work for a government that was launching a war of aggression against Iraq, an almost defenceless nation, a war that claimed the lives of hundreds of thousands of innocent people, does not disqualify the author from offering guidance on how humanity should wish to live. The Eden 2.0 sought by the liberal bourgeois is one that remains free to launch devastating attacks on weak nations which are refusing to toe the liberal line. At least, those crimes do not feature as an issue of any relevance to the points the author wishes to make.

Shortly after beginning his post, the author found himself seconded to the UN Secretary-General's office to help organise the UN's first head of government level climate summit. The author reports his excitement was short lived, when he realised the panel 'far from engaging with the big issues of fairness that arise in the context of environmental limits, ... would in fact barely be willing to acknowledge the existence of any natural limits in the first place.' This realisation caused the author to reflect

> 'if we're to overcome an issue as enormous as climate change, then we need to look far beyond policymakers and pie charts, and towards the kind of mass movements that have in the past ended slavery, fought for civil rights, or won the write-off of billions of dollars of third world debt.'

These examples are included as legitimate analogies, because all were achieved within the guardrails of bourgeois liberalism. Whether they are accurate and useful corollaries for climate change is not justified, merely assumed.

The title of Evans' book offers the prospect of retrieving the unity lost when humanity was ejected from the garden of Eden. Except the "2.0" descriptor gives a hint of the technological dimensions of the utopia the author has in mind for us. The book holds out the admirable hope that we might bring concentrations of CO_2 in the air back down from their current level of 400 parts per million (ppm) to the 280 ppm at which they stood before the industrial revolution. What, asks Evans,

> 'if we perceived our primary job for the next few centuries as being to nurse the earth back to health? The fact that this idea seems such an impossibly distant notion is *exactly* why we need stories of real power – myths – to help us find our way there.'

Those myths can't simply be imposed on people from the top down. We must all collectively and voluntarily agree to help build the kind of world people like Alex Evans want. 'But in the case of a challenge like climate change, we can't simply be passive recipients of myths offered to us by leaders. Any assumption of having "command and control" over campaign messaging strategy would be a clear non-starter.' Evans explains that we shouldn't be disagreeing with him because we 'need stories that unite rather than divide us'. That we can be united by stories, regardless of material reality, is the kind of wishful thinking required to sustain the liberal illusion that climate change can be tackled without social conflict. It is difficult to understand how such stories can be described as "different" from the current mainstream of liberal discourse.

Nonetheless, Evans does divide rather than unite by dismissing certain types of stories

> 'Almost unnoticed, we have slowly lost the old stories that used to bring us together and help us to make sense of the world. In their place, new "anti-myths" are flourishing –... that the world is heading rapidly towards environmental collapse and there's not a thing we can do about it.'

Here, again we find ourselves lost in the contradiction and confusion of liberal thought, with talk of a myth that can speak to a lot of us, but not all of us. Some groups will need their own myths. How these different myths will co-exist is not certain. We are certainly already familiar with what happens to cultures and people who have a culture different from Western liberalism. They get destroyed.

Early in the book Evans makes clear that we need to avoid the blame game, in the name of finding this unity.

> '"Enemy narratives" that tell us why climate change is all the fault of Exxon or Saudi Arabia won't cut it.'"

One must assume the need to avoid blame also extends to not blaming the government that attacked Iraq for any wrongdoing. There is, however, one group of people with whom we mustn't seek unity, one group who will never be let into Eden 2.0- those who don't want to quietly accept the liberal way of responding to climate change.

> 'So one myth that clearly *won't* help us to confront climate change is the myth of collapse: the idea that we're inevitably heading for a massive ecological crash, and that the best guide to our shared future can be found within the pages of post-apocalyptic fiction. Unfortunately, the collapse myth is becoming all too prevalent.'

We must devolve power to the people, as long as they are thinking the right things, the things we are telling them to think.

> 'It's already clear, after all, that climate change will make for an extremely turbulent – and at points downright scary – ride over the coming decades. And that's why the collapse myth is so dangerous: because of the risk that, as the reality of climate change starts to sink in, it becomes a self-fulfilling prophecy.'

It is difficult to know what is meant by "turbulent" and "scary" and whether that is turbulent and scary for the West, and something worse for poor people who live a long way away, but then, they are used to suffering at the hands of the West so we shouldn't concern ourselves too much. I am sure if "turbulent" and "scary" was all that we faced then climate change wouldn't be such a problem – you get worse on a roller coaster. No, the real danger is people start stepping outside the liberal guardrails and begin paying attention to the climate science.

'Of course, it might be that a couple of centuries from now, hindsight will show that the preppers' gloomy prognoses were right. But if that does prove to be the case, it may well be in large part the *result* of collapse myths: because enough people believed them, concluded that there was nothing we could do in the face of climate change, and did nothing.'

Evans' claim that we can and should avoid the blame game collapses under its own contradictions before we even close the pages of his book. We have an enemy – those who don't believe in liberal responses to climate change. They are the ones who are ultimately responsible for the failure to solve climate change.

What, asks the author, does Eden 2.0 look like? Pretty much like today, minus the greenhouse gas emissions – a liberal capitalist technocratic democracy, where everyone just wants to work for the owners of capital.

'To start with, we'll have moved to a zero carbon economy…we'll have started taking carbon out of the air and putting it back in the ground…This will be a world in which our economy has stopped being in tension with environmental health, and has instead become the most powerful mechanism we have for restoring it, with tax systems overhauled away from taxing things we want (like work) and towards things we want to discourage (like pollution or resource use), with the effect that companies increasingly com-pete on the basis of how small their environmental impact is. And all this will be happening in tandem with social goals, not in tension with them. By 2030….inequality will be far lower than it is today.'

The future is looking very positive indeed.

The use of the "2.0" adjunct – pulled straight from the digital industries – tells us that this will be essentially a technological utopia. Alongside the standard refer-ence to something called "wellbeing", Evans gets tangled up in whether or not we need a common uniting myth or many different ones.

'We need a new generation of twenty-first century myths that can explain… who "we" are. Specifically, we need myths that get us to think of ourselves as part of a *larger us*; that situate us in a *longer now*, with much greater aware-ness of the deep past and deep future…. In a world of 7 billion people, there's clearly no single myth or set of myths that will work for all of us.'

There are no class conflicts to be fought out, no more wars of aggression, just the peaceful negotiating of respective myths that thankfully account for the ecological limits posed by climate change. Having laid out an argument that the right way of being in the world is the Western way, Evans tries to retrace his steps.

> 'The terms on which climate policy will play out over the next two decades have now been set. This is to be one of the most technocratic agendas the world has ever seen. Climate change will be owned by a "priesthood" of experts, with its own language, rituals, gatherings, and assumptions – and, above all, abbreviations. It's a thoroughly insider game, and the ragtag bands of activists outside the summits are a charming irrelevance. As for the public, their job is to listen to the experts and then remember to turn out the lights. It's certainly not to *participate*, much less wield *power*.'

This will all be possible because the relationship between ordinary people and technology will – by undetermined means – be overturned.

> 'What's missing in this story of being sucked towards some kind of techno-logical cliff edge is the idea of people as the prime actors, or the sense that we have agency and choices over how the future unfolds, at what speed, and for whose benefit.'

Such words sound very much like the promises made by technocrats responsible for our current impotence in the face of technological change. Though these concerns are not original – in fact, they have become something of a cliche in social science discussion of climate policy – it is a welcome recognition of the challenges ahead. Given the author's name dropping in the earlier passages in the book, one wants to take his plea for distributing power to the mob seriously. But, it is more likely another expression of the confused, chaotic and contradictory nature of the bourgeois liberal ideology. And so it is that we shortly after encounter passages praising life in a technocracy, where the victory of bourgeois liberalism is complete

> 'We're poised right on the cusp of a genuinely global "us" – a global social media network, a global library of knowledge, a global economy, global governance institutions, a global sense of who we are.'

6.5   What we think about when we try not to think about global warming: towards a new psychology of climate action

Stoknes begins by telling his readers

> 'I had grown up in a family owned smelly fish factory on Norway's gorgeous western fjord coast and later exercised the entrepreneurial and curious genes spawned there in the worlds of green tech and plasma physics.'

(xvi)

Much like Evans, Stoknes 'background is firmly upper-middle class and bourgeois. Here, Stoknes goes a step further than Evans, and is clear that the vision he offers is a product of the social and historical conditions of his upbringing. For Stoknes, climate change can be solved, and the solutions already exist. His upbringing allows him to see

> 'We have the solutions we need to fix climate change: from radical energy efficiency to renewable energy, better education for women, reforestation and carbon capture.'
>
> (xviii)

There is no need for fundamental economic or political transformation in order to deal with climate change. In arguing against the claim 'people, capitalism and democracy are too short-term to tackle the critical long-term climate issue' (xvii), Stoknes instead suggests the problem is that 'As long as there are few opportunities for simple and easy climate-friendly behaviour, and the message stirs feelings of fear and guilt then people will reject the messages' (64). 'We just need more and more of us, as consumers, to keep shifting gradually toward voting with our wallets for greener purchases' (92).

Stoknes agrees with Evans. The real problem is people not having sufficient desire for the world that people like Stoknes and Evans want

> 'Climate scientists could partner with psychologists, sociologists, artists and social scientists to communicate the science through visualisation, frames, and stories that foster action and hope, rather than despair and denial.'
>
> (Xix)

'We should begin to talk about climate change in terms of opportunity, and stop framing climate change in terms of loss' (111–112). Why, after all, should we let a crisis go to waste? By viewing climate action as an investment in future profitability and competitiveness, 'We get better growth *and* better climate' (119).

For Stoknes, we need 'stories that emphasise well-being, social justice and generosity as the new wealth' (139). Quite what any of this means in practice is not immediately clear, but then two pages later, all is made clear. Stoknes dreams of a 2050 with an astonishing diversity of cultures (141), but goes on to describe just one culture, a Western, white middle class liberal one, as the ideal type

> 'There is flirting, gossip, philosophical cafes, street theatre, peaceful protest marches, farmers markets, marathons and rock concerts a plenty. The cars hum quietly around…beer and barbeques. Markets and trade are vibrant.'
>
> (141)

The author proposes a remarkably stable set of social and economic conditions in the 2°C future. There are no climate impacts, no conflict. Everything is peaceful, entrepreneurs are trading, everyone is experiencing "wellbeing." All will all be possible in a world that has warmed by 2°C, regardless of scientific projections to the contrary: 'We must at least halve emissions by 2050 to have a fair chance of avoiding too much runaway climate disruption. This could allow us to meet the famous 2 degrees Celsius goal' (154).

Echoing the point made earlier in this chapter, Stoknes argues we must integrate science with storytelling (148) to ensure acceptance. And we must maintain our faith in the possibility of salvation through liberalism.

> 'The apocalypse story only reinforces the old barriers, while stories of green growth…offer us a way round the most deeply seated barriers in the minds and hearts of modern citizens. We will willingly shoulder the burden of acting long term if we have…some kind of grand stories about where we want to go that give us a sense of common purpose.'
>
> (150)

All that matters is that those stories are the stories Stoknes likes.

6.6   Conclusion

There is a hope that stories can help bring about a smooth and painless transition to a new net zero future. The dominant stories of our net zero future portray a world free from conflict and climate impacts. The stories of our net zero future are generally written by the liberal bourgeois, and reflect the interests, privileges and hopes of this class. Liberal claims that stories can change the world contradict the assumptions of historical materialism, wherein it is changes in the material conditions of existence, and the mountain of contradictions and crises therein, that drive historical change. Climate change is such a material force, whose increasingly inescapable crises threaten to bring about profound change. Unfortunately, the unrealistic wish fulfilment stories of a liberal paradise that currently dominate our visions of the future are ill-placed to guide us through the impending upheavals.

Note

1 I share the myopia that comes with being a white Western male. My perspective as someone who comes from an impoverished working class background at least gives me some distance on some of the privileges shaping the ideas the two authors promote in their books.

References

Anfinson, K. (2018) 'How to tell the truth about climate change'. *Environmental Politics*, Vol. 27, No. 2, 209–227.

Bäckstrand, K., and Lövbrand, E. (2019). The road to Paris: contending climate governance discourses in the post-Copenhagen era. *Journal of Environmental Policy & Planning*, Vol. 21, No. 5, 519–532.

Baudrillard, J. (1968). 'The system of objects', in Poster, M (ed). *Jean Baudrillard. Selected Writings*. Cambridge: Polity Press, pp. 10–28.

Bernauer, T. and McGrath, L. (2016). 'Simple reframing unlikely to boost public support for climate policy'. *Nature Climate Change,* Vol. 6, 680–683.

Bray, D., and Shackley, S. (2004). The Social Simulation of The Public Perceptions of Weather Events and their Effect upon the Development of Belief in Anthropogenic Climate Change, Tyndall Centre Working Paper 58.

Burr, V. (1998). 'Overview: realism, relativism, social constructionism, discourse', in Parker, I. (ed). *Social Constructionism, Discourse and Realism.* London: Sage, pp. 11–13.

Cass, L. (2007). 'Measuring the domestic salience of international environmental norms: climate change norms in American, German and British climate policy debates', in Pettenger, M. (ed). *The Social Construction of Climate Change: Power, Knowledge, Norms, Discourses.* Aldershot: Ashgate, pp. 23–50.

Cook, D. (2018). *Adorno, Foucault and the Critique of the West.* London: Verso.

Crenson, M. (1971). *The Un-Politics of Air Pollution. A Study of Non-Decisionmaking in the Cities.* Baltimore, MD: Johns Hopkins University Press.

Dryzek, J., Downes, D., Hunold, C. et al. (2003). *Green States and Social Movements.* Oxford: Oxford University Press.

Evans, A. (2016). *Eden 2.0: Climate Change and the Search for a 21st Century Myth.* EBook

Ghosh, A. (2016). *The Great Derangement – Climate Change and the Unthinkable.* Chicago, IL: Chicago University Press.

Goodson, I.F. and Scherto, R.G. (2011). 'The narrative turn in social research'. *Counterpoints*, vol. 386, 17–33. Retrieved from http://www.jstor.org/stable/42981362

Ingebritsen, C. (2002). 'Norm entrepreneurs: Scandinavia's role in world politics'. *Cooperation and Conflict*, Vol. 37, No. 1, 11–23.

Jackson, L. (1994). *The Dematerialisation of Karl Marx. Literature and Marxist Theory.* London: Longman.

Kingsnorth, P. and Hine, D. (2009). *The Dark Mountain Manifesto.* Retrieved from https://dark-mountain.net/about/manifesto/ Accessed 20/05/2023.

Li, N. and Su, L.Y.-F. (2018). "Message framing and climate change communication: A meta analytical review'. *Journal of Applied Communications*, Vol. 102, No. 3

Mair, P. (2019). *On Parties, Party Systems and Democracy: Selected Writings of Peter Mair.* Colchester: ECPR Press.

McCright, A.M., Charters, M., Dentzman, K. and Dietz, T. (2016). 'Examining the effectiveness of climate change frames in the face of a climate change denial counter-frame'. *Topics in Cognitive Science*, Vol. 8, 76–97.

Newell, P. (2000). *Climate for Change: Non-State Actors and the Global Politics of the Greenhouse.* Cambridge, MA: Cambridge University Press.

Nisbet, E., Hart, P., Myers, T. and Ellithorpe, M. (2013). 'Attitude change in competitive framing environments? Open-/Closed-mindedness, framing effects, and climate change'. *Journal of Communication,* Vol. 63, No. 4, 766–785.

Peckham, M. (1995). *Romanticism and Ideology.* London: Wesleyan University Press.

Prawer, S. (2011). *Karl Marx and World Literature.* London: Verso.

Ramanzanglo, C. (1993). *Up Against Foucault. Explorations of Some Tensions between Foucault and Feminism.* London: Routledge.

Roberts, J. (2004). *Environmental Policy,* London: Routledge.

Savage, L. (2017). 'How liberals fell in love with the west wing'. *Current Affairs.* Retrieved from https://www.currentaffairs.org/2017/04/how-liberals-fell-in-love-with-the-west-wing

Schmidt, V. (2008). 'Discursive Institutionalism: The Explanatory Power of Ideas and Discourse.' *Annual Review of Political Science* Vol. 11, 303–326.

Sebald, W.G. (2014). *A Place in the Country.* London: Penguin Books.

Shaw, C. (2016). *The Two Degrees Dangerous Limit for Climate Change. Public Understanding and Decision Making.* Abingdon: Routledge.

Spark, C. (2001). *Hunting Captain Ahab: Psychological Warfare and the Melville Revival.* Kent, OH: Kent State University Press.

Stoknes, P. (2015). *What We Think about When We Try Not to Think about Global Warming: Toward a New Psychology of Climate Action.* Vermont: Chelsea Green Publishing.

Washer, P., and Joffe, H. (2006). 'The "hospital superbug". Social representations of MRSA'. *Social Science and Medicine*, Vol. 63, No. 8, 2141–2152.

Wells, A. (1987). 'Social representations and the world of science'. *Journal for the Theory of Social Behaviour*, Vol. 17, No. 4, 433–445.

White, C. (2003). *The Middle Mind. Why Consumer Culture Is Turning Us into the Living Dead.* London: Penguin.

Whyte, H. (1973). *Metahistory: The Historical Imagination in Nineteenth Century Europe.* Baltimore, MD: John Hopkins University Press.

Willig, C. (1998). 'Social constructionism and revolutionary socialism- a contradiction in terms'? in Parker, I. (ed). *Social Constructionism, Discourse, and Realism. New York:* Sage Publications Ltd., pp. 91–104.

7 Maybe tomorrow

7.1 Interview methodology

This chapter presents the results from an analysis of 14 interviews carried out in the summer and autumn of 2022. The interviewees were selected on the basis of their climate change expertise, and because their work in one way or another speaks to the question of what kind of society is needed to avoid ecological disaster. The other criteria for selection were that, whether through advocacy, teaching or some other aspect of their role, they were talking about climate policy and climate politics, and hence in some way have the opportunity to shape the discourse.

The interviewees were drawn from the fields of academia, policy and advocacy. They were known to me through my own work and long-standing interest in climate politics. Interviewees were predominantly from Europe, Australia and North America. This geographic focus was partly about accessibility (requests to speak to climate experts from the global South were largely unsuccessful). But it was also because the focus of this work is to understand how experts in the global North respond to calls for the 'fundamental changes' to our way of being that the IPCC say are required. If resources and time had allowed, it would have been desirable to compare answers to the same questions from actors in the global North with those from the global South.

I am deeply grateful to the interviewees for their time and candour. I am also profoundly grateful for the work these people are doing to build a movement for change. I hope this book in some way does honour to the change they are building under such difficult circumstances. Whilst this book is a critique of liberalism's failings on climate change, the analysis in this chapter is not a critique of the views held by the interviewees or the interviewee's work. The interviewees are representatives of, or drawn from, the class Gramsci described as 'organic intellectuals'. This is the class that, for the purposes of this book, is considered as shaping the boundaries of the climate debate, mediating between the people and power. The class of organic intellectuals includes scholars and experts such as those interviewed here, but extends to technicians, economists and managers. Organic intellectuals are essential to both the maintenance and also the transformation of social relations. The ability of the class of organic intellectuals to bring about social transformation is to a large extent shaped by institutional norms that constrain and

DOI: 10.4324/9780429463488-8

define the activities of people working in these fields. The interviews show that many are, in their professional capacity, struggling to fully articulate what they know and feel should be done. As one interviewee said, regarding the inclusion of calls for fundamental change contained in the latest IPCC report 'Someone must have lost their job when they realised that got through.' (M1).

The interview questions centred around the following IPCC statement mentioned above, namely

> Targeting a climate resilient, sustainable world involves fundamental changes to how society functions, including changes to underlying values, worldviews, ideologies, social structures, political and economic systems, and power relationships.

The interviewees were asked

- The extent to which they agreed with this statement?
- What fundamental changes they thought needed to happen?
- Who would drive this change?
- What this new world would look like?

Each interview lasted about 30 minutes.

This is a thematic analysis. I do not focus on clustering responses by the respondent's role. The number of interviewees is not sufficient to be confident that any patterns that emerge are the result of a particular role (e.g. sociologist, campaigner, scientist, engineer or communications expert) or demographic characteristic (e.g. gender). Second, to protect the anonymity of the interviewees I am removing any identifying characteristics, which means not tying detailed accounts of a person's responses to their role.

The answers were, of course, a function of the questions asked and the people interviewed. Different questions and different interviewees would undoubtedly have revealed different perspectives on the same issues. These interview answers don't necessarily reveal some eternal truth regarding laws of social relations. They are a snapshot, and the answers are given in the context of a number of influences and issues. What the answers seem to have in common is an orientation away from any revolutionary social and political change as intimated by the IPCC statement, and towards voluntarism, quietism and smallism. This is an inevitable response from citizens of an atomised society to the overwhelming scale and seeming impossibility of bringing about the kinds of changes the IPCC describes as necessary. Ultimately in liberal societies we are all facing these questions from a position of social isolation, perhaps more so for largely middle class career-oriented experts. Few of us, whether climate expert, climate radical, or a citizen going about their daily life and work far removed from discussions of climate policy, can see a way to deliver the changes the IPCC say are needed to avert disaster. In my day job, I have sat in many focus groups throughout the UK, with largely working-class people selected purely on the basis of their (normally centre-right) values, with no expertise in climate change. Having spoken to not well-educated working-class people and highly educated climate experts, I cannot identify any meaningful difference in the ability to imagine a alternative future, or a route out of this mess. No one has the answer, there is no blueprint for salvation, no one is in control.

7.2 Results from the interview analysis

7.2.1 *Freely choosing a future of fewer freedoms*

The theme of consumption emerged spontaneously and unbidden as a response to discussions about where the kinds of changes described by the IPCC might come from, who might initiate and drive such changes. This pattern in the answers reveals an almost pre-theoretical grounding in our shared understanding of individualism as something expressed primarily through purchasing behaviour. Freedom in liberal societies is primarily the freedom to choose what to buy and where to travel, with the only limit being access to capital. This behaviour needs to change, and is an important locus of climate action.

> *It's too simplistic to say any focus on personal overconsumption is just a delay tactic or a distraction by big fossil. We won't have enough clean energy to go around, unless we all have a reasonable lifestyle, basically.*
>
> (F1)

> *I don't see any massive reductions in greenhouse gas emissions at the scale that we need without changes to how we think about consumption, how we think about production and reproduction. And then there's this economic dimension of completely reforming how and what we go about purchasing.*
>
> (F2)

Ultimately, change will have to be brought about on a voluntary basis, the new behaviours must be freely chosen.

> *I think you have to give people a way to change. Choices are linked to ego, a sense of self, a sense of who they want to be in a community, a sense of belonging and all of that. To shift behaviour, there needs to be a route to do that. That isn't like whipping people all the way along. And so in more concrete terms, that means making it more expensive to own that other thing.*
>
> (F2)

One interviewee highlighted how in her work on mobility and climate change she feels unable to pursue fully the implications of the climate science, to suggest limits be placed on people's freedom of movement.

> *Instant mobility is basically a function of capitalist society and a function of the fact that we've all been reduced to these individual economic agents*

and consumers, who have this fundamental right to a holiday. Even the most radical people are really nervous about challenging the idea of everybody having their two weeks in the sun.

(F3)

It wasn't clear where the decision to choose to consume less would come from, what the trigger would be. One interviewee saw the climate problem as driven primarily by the middle classes of the West trying to impose their values on subaltern groups who have much lower levels of consumption.

We always want ever more, we want more, more stuff. We think we're worth it as well, but also, we are often dissatisfied. So, we academics and campaigners are very unlike those other communities.

(M1)

In a similar vein, discussions of choice required imagining a citizenry with the privilege and capacity for choice making enjoyed by the middle classes but perhaps not by less well-off and less well-educated parts of society

We, as individuals, have opportunities to respond as well. And that's in these five roles, not just as consumers, but as role models; and how we influence others as investors; and how we use our money in banks and the financial system; as citizens, and how we engage politically; and as professionals, as a part of organisations or collective action, through climate movements. Those are the levers we have to pull.

(F1)

7.2.2   The individual's role in creating a system of fossil fuel-free exploitation

The quote above highlights the difficulty of identifying how individual change creates large-scale change. One idea is that sufficient individuals changing their behaviour will lead to system change, while the system, spontaneously changing concurrent to this shift in individual behaviour, will create the conditions for the wider adoption of these behaviours.

There was no universal agreement on the role of the individual in driving this change. Noting that 'We don't really like to change our mind' (M3), interviewees highlighted the need to look outside ourselves for the change to begin. Some interviewees argued the system must change first, before people will.

This thing about just go out and strive to achieve for yourself and you're the centre of everything. I don't think this is intrinsic in any way. I don't think it's even a long-standing aspect of our societies. I think it's been manufactured.

(M5)

I guess at the heart of liberalism is the sort of Enlightenment, Cartesian rational individual, a focus on the individual, a focus on rationality and I think there are limits to that. The worm in the liberalism apple is the limited individual framing.

(M3)

People don't have the brain space in the day-to-day to step back and consider it. Where do you find space to consider how you're going to disrupt that system? We're very easily adaptable and we know a situation that we're in and we find it hard to picture things outside of that. Like with Covid we adapted very quickly but also afterwards we've very quickly forgotten what it was like.

(F4)

Essentially, people think first we have to change the attitudes and then the behaviour will follow. What I vaguely understand about behavioural science is actually, it's often the other way around. And there's also this model of first, let's agree on the science and then the policy will follow. That's not always the way that human beings behave or things are decided.

(M3)

If you're born into a rich family and go to Eton, you'll probably have quite different values than if you're born in a slum in Cape Town. These things come from our circumstances. I guess this is the kind of the hardest stuff to change, is the kind of really deep-rooted stuff.

(M6)

Yet, in the absence of any sense of how and why the system might voluntarily destroy itself to make way for a less destructive alternative, interviewees returned to the role of the individual. The task is to identify a way for the individual to join the movement for change whilst, at the same time, the system goes about the work of changing itself. There is a well-known optical illusion drawing of a woman's face. The picture shows both a young woman turning her head away from the viewer, or an old lady with her head wrapped in a scarf, depending on how one looks at the picture. In fact, one feels one can see both images at once in the same picture almost at the same time, both young and old. In a similar way, it is possible to comprehend system change and individual change occurring at the same time, both dependent on and at the same time independent of each other.

The issue is that you have to get people on board at a point in that journey that makes them feel as though they're having an impact and therefore they're more willing to engage further with things that will cause system change. Even those small things can feed into that system change. And so, I don't think that people would necessarily deeply disagree with the fact that you need system change at all.

(F4)

Yes, that we haven't got another hope, right? So I have to say, I have to say that. Yes (the people can be the engine of change). Because otherwise, I just go back to what I used to do for a living. And you know, you need two things, you need that flashpoint to happen. You need people to have that awareness of what it is and that it's a problem and it's and it affects a big enough number of people, right? But the second thing is you need the structures that are there to take that energy and that momentum and do something with it.

(F5)

I'm tired of this oversimplified individual versus collective action idea. And multiple things are true at the same time. I mean it's certainly true that no individual single-handedly has the power to stop climate change, but it's also true that climate change won't stop until the whole world collectively has stopped producing and consuming fossil fuels, and stopped emitting carbon to the atmosphere. Some climate activists say that the only action that matters is collective action. Most people do not know what they mean by that or what that looks like. And most people are not politically engaged. I mean, most people don't support organisations with their money or time that are working for climate action. So building that bridge, from where people are now to where they need to be, you need to have steps along the way that can bring more people with you.

(F1)

Change can only come from individuals, but then the point being what is an individual? The problem with the word individual is – and I think this is why talk about behaviour and individual behaviour grates with some people – is it immediately implies an atomized view of society, where we all just go round consuming and it's all about our carbon footprint. I think both in academia and activist circles it seems like you have to either occupy one or another one, one of two extreme positions sometimes. Either everything's down to individual behaviour and consumerism and change the prices, or give people the right information and nudge them. Or people are, if not a dead mass, then kind of just a big blob of zombie. For me, I think it's the between that we don't focus enough attention on and think about. In a social movement like Fridays For The Future that exists because of lots of individuals being part of a bigger thing. But ultimately, in the West particularly it feels like we've forgotten we are relational, we exist with and through other people. If the question is, can the change needed be driven by people through their networks and collectively and so on, the answer is yes.

(M6)

Change will come primarily from people's movements – indigenous peoples, other local communities, civil society groups; but governments in some cases will also play a role. I am not optimistic about any substantial corporate involvement in driving fundamental changes – primarily because they would have to dissolve themselves, at least in the form they are now.

(M8)

So, whilst fundamental change might be expected in some fields of political activism to result in an overthrow of existing institutional structures, the people interviewed for this book frequently expressed a need to retain many of the features of liberal market economics and governance to deliver climate goals.

I think there is a trap in being obstructive, focusing on everything that's not good enough or may just be nice green packaging on a thing that's not going to get us there. I think we need to work all the levers, move in all the directions and hopefully things will get there right? And I think those are things that climate finance, and some of the green growth stuff, the STEM (science and technology – oriented solutions) stuff like those are all partial solutions. I think they are going to maybe help us make headway and I wouldn't say I wouldn't go maybe to the other extreme to think that we need to undermine or throw out any efforts in those directions.

(M2)

7.2.3 Searching for mushrooms

Interviewees demonstrated great faith in individuals as agents of change. Hope in the individual is maintained in the absence of evidence that large numbers of people are willing to change both their behaviour and the society they live in. This hope lives through the analogy of mushrooms. Though the fruit of people's awakening is not yet visible, there is a sense that out of sight something is growing and about to break through the ground into the light of day, following the rain of enlightenment.

If you are talking about more superficial attitudes, I don't know, wanting to have five holidays a year for example, that's a pretty recent phenomenon. It's malleable, we can shift that stuff. The reason why fracking is not politically possible is because of people doing activism, is not because some of what engineers said.

(M6)

I'm by nature quite an optimist, actually. The insights emerging out of the behavioural sciences about human nature gives me great hope because they open up the possibility that we may begin to more truly understand who we are as human beings and to dismantle shibboleths about us being sort of individual rational, logical creatures.

(M3)

The other big problem I see is actually people not believing that what they do matters. And they do actually have so much untapped power. I think there's so much that can be done. Because you really need this critical groundswell of enough people doing enough things that you can reach a social tipping point, where the kind of changes that we need are possible. We know from social science research, that people need to believe that their actions matter. They need to have efficacy before they will take action, which is pretty reasonable, but that's kind of a catch-22, right? Because taking that action, you might not see the visible results of it in the short term or maybe ever. I mean, you don't necessarily know the impact that your actions will have on others or in laying the groundwork for someone else to do something. And you might never know that person but they actually make the critical difference. So I think that's really a challenge to inspire people or convince people that what they do does really matter.

(F1)

Someone said it's like mushrooms. You can go out in your garden, you're like, oh, you know, the lawn's looking a bit shit. And then the next day, green. Now, there is always something going on. There's always a group that wants things to be better. The thing that hasn't happened yet is there's not been one collective demand that pulls it together, right? And that will happen. Fifteen years down the line, someone will be writing a book claiming that it was a very clear, linear progression present. Martin Luther King turns up in the middle of a boycott. He just was the face of it and then behind the scenes, you have all these unknown heroes. That people don't even know the names of them now, who did all the joining up and did all the fundraising and did all that, you know, all that stuff that doesn't get written about. And so, it's really difficult when you're in the midst of something to know what's going to happen. But I would say, like there are pockets of people, educating themselves that are pockets of people taking action, there are pockets of people holding people to account, But there hasn't been that collective, like flashpoint yet. And I think with climate change, it can be difficult, right? Because it's overwhelming. And how do you make the demand tangible? I do genuinely believe that triggering a social movement will drive this rapid radical change.

(F5)

So I do see lots of places where change is possible though. I do want to point out the communal and communitarian roots of many conservative movements. I think that is there and certainly the work I've done on community energy, it's not simply urban elites versus rural workers. You know, there are a lot of groups in rural areas finding quite a bit of meaning in a lot of communal activities now. It's excluding others in other ways and it gets complicated, but I see space for coming together around that.

(F2)

I think there's a lot of reason for hope. There's a paper from Science that says it takes 25% of the population adopting a new social norm for the rest to follow and research would suggest that already there's a huge number of people who feel willing to address this. But it's not made visible because they don't know how to make it visible. We constantly have that argument, what if we had leaders who took these bold steps? I always use that quote 'the

heroes of tomorrow are the radicals of today' kind of thing. Then other peo-
ple will be so grateful that they've been bold enough to take those steps. But
some doesn't see that because it's not visible to them, the voting electorate's
actual underlying opinions and fears etc, are not visible.

(F4)

It was clear then that change is possible, but the change was not likely to be the
fundamental reordering of our values and societies that organisations such as the
IPCC have begun calling for. There is little desire or energy for such change in
society, and in any case the conditions for a coordinated move to a more secure and
stable future have now gone.

I'm not convinced. and I wish though that we were a point where people are
going to start going out on the streets and saying, we want X, Y, Z, that we're
going to get a Winter Palace storming anytime soon.

(F5)

I just think It's whatever happens, we're not in control of it anymore. You know,
I feel like we're so far behind where we could be if we actually confronted it
that I can't even think about it because it terrifies me so much. I feel like I'm
having a therapy session here. It's not going to be in a controlled manner, that
the changes are going to happen and we will not be in control of them.

(F3)

Okay. Yeah, fine. If you want to get involved, that's great. Let's make the thing
as open as possible so that people can get involved in the politics and the
rest of it... But I don't see why you have to be in it... we can't all be experts
in everything, all the social science. 'You have to bring everyone on board'.
What, me? Yeah, awesome. But actually a lot of my friends live bloody good
lives, they enjoy their life. People like me have this need to progress up some
ladder. They don't have that most. Well most of them don't. A mate of mine
drives ambulances. He wants nothing other than to drive ambulances and
enjoy time with his family, enjoy quite a lot of beers and do loads of cycling.

(M1)

7.2.4 *Keep your head down*

Whilst we are awaiting the lightning rod, the leader, the vision to unite and galvanise
us, people are having to get on with the day-to-day business of earning a living,
which often stands in the way of pushing forward the work needed to usher forth
the mushrooms under the ground. Instead, we are focused on sustaining our ability
to earn money, whilst identifying the small wins.

None of us have control over the outcome. I mean none of us can single-
handedly say 'Yes we stopped climate change' or 'No we didn't'. So many

things have to line up. So that's part of the reason I'm more focused on the process than the outcome. I don't have this utopia really that I dream about, that gets me out of bed. It's more like, this is the work that needs to be done and these are some steps on the path to get there. I don't know, I guess it's also maybe a bit of a risk because there won't just be one utopia if we actually succeed beyond all our wildest dreams; there has to be room for everybody's world there, you know. And so I guess I'm not focused on the specifics of, you know, what do buildings look like? Or how do people get around... and that's where maybe climate fiction has been helpful for me to envision and paint different realities, and reinforces for me this idea that it's helpful to have this diversity of views and not just think that it will be one way.

(F1)

I think there are definitely choices of language that I've made in my own sort of experience of writing that book that don't reflect really what I think.

(F3)

As a citizen, I question those climate targets. As an academic. I don't. I say right? Oh fine, you tell me you want to stick to 1.5? I'll tell you what, I can tell you from the science if you like and my understanding of these issues and I don't care whether you like, it, or don't like it, I don't care if you agree or disagree with it. And so that's what I feel, my job is not to question the 1.5 or 2, not as an academic.

(M1)

I feel trapped within the current system, right? And you know, as an academic and who you know, from time to time contributes to broader debates and but also, to the policymaking system I feel there's certainly a tension between what I think would be required to deliver that kind of transition or transformation and what I think the public or the policy system is capable of.

(M4)

But there are multiple forks in the road. Nothing is inevitable. Apart from a few science fiction writers, most people wouldn't have imagined we'd be living as we are today now in 2022. And I don't reckon apart from the people who like Margaret Atwood or dystopian, Mad Max people, you know, there aren't many people who've imagined a century, hence, how we might be living if we have dealt with this stuff. And all we can do is try and do things that we think are pushing the right direction and avoid things we think are pushing in the wrong direction. Exploring civil disobedience and whatnots and just stop oil and so on. There's all sorts of ways where that might have been, and probably is, problematic, and maybe switches some people off and maybe is unhelpful, but broadly speaking as a kind of blunt instrument is that stuff helping push things in the right direction? I think yes.

(M6)

7.2.5   *Substituting politics with science and technology*

The grand and radical visions of radical social change such as that from the IPCC offer a more challenging route out of the crisis, compared to the relatively painless promise offered by technological responses. The sense of progression and improvement is more tangible and less messy than trying to change a society's values.

> *Yes, I would agree with that stuff in the IPCC statement but then it comes down to what all those things mean, doesn't it. We talk about radical, listen, radical that. But it begs the question. What does that actually mean?*
>
> (M6)

> *We tend to forget that a bit I think, because there are technical answers and I think as a society, we feel more comfortable with technical processes because there's a theory, there's a formula, you can follow and reach a desired endpoint there. They are predictable. You know, if you put in place policy A you might be more likely to get outcome B.*
>
> (M5)

> *I think also for me the Sustainable Development Goals have the social dimensions a bit more woven throughout, which give a nice counter balance if we're talking to policymakers, decision makers, challenging what I feel like has become a very STEM captured, green growth, clean growth captured kind of language in the official circles of debate in places like the COP meetings. And I look at the Sustainable Development Goals it's like, okay, there's this thing that has some buy-in at least at the upper level of global conversation that we can maybe use to help, provoke some change or some direction.*
>
> (M2)

> *Going back to that statement of values and the beliefs and I think that to me is the absolutely the most challenging because that's linked with politicians and what's politics and so on.*
>
> (F6)

And yet, it is an illusion – according to an interviewee who takes actions as part of XR campaigns – to imaging science circumvents rather than simply informs value decisions.

> *I suppose the science provides us with warnings. It provides us with a very clear required route of travel. Of course, it informs all of our decisions but it doesn't determine the decisions that need to be made on that, because the decisions are ultimately our values.*
>
> (M5)

Spiritual beliefs, especially in the US, were identified as one possible counter to material values, but even then, the possibilities seemed limited. Religion provides an existing set of strong values from which to build. The problem is, science and technology has displaced spirituality as an ordering principle.

> *I have in my head an example of someone who is at the end of the spectrum, right? They've got two giant vehicles, a massive house. They're very invested in preserving a maximalist consumption lifestyle and maybe resistant to a lot of these green initiatives and things like that. But often these groups are also quite religious. And I think there is maybe some space there to talk about values and where you get your validation from in your life. I think we have to find those ins, and those are going to look different depending on that group and where it's drawing that validation from. So I think churches and community groups are one place for a lot of people even if they hold quite conservative values.*
>
> (F2)

In the absence of these ordering principles, these other strongly held values, we have no route for action and are left confused by the conflicting information we receive. Hence it becomes necessary to look outside ourselves, our own class (the bourgeois) for answers.

> *Flying is okay now because actually that's not such a big deal, you know doesn't have that many emissions so I shouldn't feel guilty any longer, and the next day, these graphs appear of how many emissions come from the private jets and this and this and this. I think we're just confused, people are confused by this conflicting information, but then I think our attitudes are not rooted in any kind of fundamental understanding of what the issues are, or even values to be honest.*
>
> (F6)

> *There's a significant – maybe 15 to 20% – indigenous population in New Zealand and their environmental worldview and ethic is very well developed and there's a lot of academic writing about it from Māori scholars as well. It is a direct challenge to individualism, neoliberalism, and many other things. It does talk about one of the biggest challenges being our relation with the planet, that we are not a separate entity from the planet. So even when you introduce yourself today as a Māori, you do so talking about where your mountain is, where your river is, where your community is, like, I am from this mountain.*
>
> (F2)

> *We need to shift from what I call the exploitation mindset to the regeneration mindset. So I characterise the exploitation mindset as based on two fundamentally flawed assumptions. First, that some humans are more valuable than others and should therefore dominate each other. So that's a failure to recognize each person's inherent value. And second, the idea that humans*

are separate from nature and should dominate and exploit nature. I mean, my gut feeling is that there are a lot of people who have values in line with regeneration. You can imagine that these more universalist and transcendental values – which are about harmony between people and nature – are widely held. Maybe those are the most in line with the kind of world that I'm talking about. But I actually think we could get there with almost any combination of values. I mean, people also have a need for security, for example, or power or these other things and those values could be met and people would be better off in a world that actually worked for everyone as well.

(F1)

Who's in that group defining the problem? Yeah, the professors, the senior NGOs, the policymakers, the entrepreneur, the barristers, that journalists, all the people that have framed the climate agenda are in the top one percent... and not surprisingly, what we don't find is an agenda that features, gender issues of equity or issues of when is enough enough? Now, obviously the whole sufficiency arguments – a few people push hard on that agenda, but they are a very very small marginalised voice.

(M1)

In the UK, for example, we have been pushed towards a more individualistic, self-interested mindset. But it's not always been there, and it doesn't always have to be there. So I think that can change. So one's values and worldviews I think it depends on how rooted, how deep breach do you go and how deeply did you think you need to go?

(M6)

7.2.6   *Talking climate*

There was recognition that people, or the media at least, are talking about climate change, but there was a general sense that these representations were often far from honest.

Five years ago, I could easily read everything that was written in the print media about climate change, because it was, you know, one story every couple of days, that was literally it. And now, you know, now it's everyday news, it's come into the mainstream.

(M4)

Overall there was a positive response to the language of the IPCC report, whilst recognising the limits of language.

So, the language has changed a lot in the last decade or so. But it's not actually changed anything in terms of what we're doing, right? It's just the PR machine has succeeded in rebranding, pretty much. What's interesting about the IPCC is that they didn't use to use this kind of language. It was very much about 'we can change little bits of things. But now, actually, you know what? The way we've built the world doesn't work except for a very few people at the cost of literally everything else. So let's have a very honest conversation about the fact that the systems don't work.' That change has been fantastic to see. The big picture stuff is changing, its language and its tone. But obviously, at the same time, you see this obsession with net zero. People have just latched on to that urgency and that language and gone, "cool, we'll sell you this version of the future and we promise to fix everything" and that's as dangerous as the version where they were just unwilling to accept reality.

(M7)

I met a brilliant behavioural scientist who had developed a concise explanation of climate change and he went on a tour of red states, to try and persuade people. And I said to him, okay, so you may get these audiences to understand and accept the science. Is this going to change who they vote for, what they want? He said, "I've got no idea. That's not my job". And ever since then I've been noticing that many scientists are focused on changing understanding of the facts rather than changes in behaviours.

(M3)

I think this is part of what we're facing now, this very reductionist narrative of the problem. There have been many really, really influential social scientists who've laid out clearly what the problems are, why we need to better understand human behaviour, why we need more social science in the decision-making process, and it still gets side-lined. I think this is part of the problem. This is something I talk a lot about, that the knowledge that we are using to develop climate knowledge is still so dominated by the natural sciences and they just don't integrate the society, this complexity in the same way and therefore we end up with a shopping list of the kinds of options and then that means that we take action. Should we maybe allow for a little bit more divergent kind of epistemological effects and actually give space for that? Everything from indigenous knowledge, traditional knowledge but not just kind of this sort of formal alternative, but actually like all of the different perspectives. I have a lot more respect for a perspective from somebody coming from India, from Bangladesh than from the media or whatever. We've framed the whole problem wrong and I think that's why we're kind of now where we are.

(F6)

There's very little interest in effective just transitions, decarbonization, those kind of conversations. The conversation is all very disingenuous. Well, our offshore oil is going to have a lower carbon footprint at the extraction end. So we should definitely pursue it.

(F2)

People know climate change, they experience it, they know the droughts are longer, they know that all over Africa. Like the weather is changing. But as for terms like 'climate change', in lots of languages, climate change doesn't exist as a word or exists in different forms and so people, they have lived experiences that climate change is real, but they use different labels for it.

(M7)

7.2.7   *So much to do, such little time*

The possibility of new norms emerging from individuals in a society which structurally supports the reproduction of existing norms was seen as an insurmountable barrier for some interviewees.

Some sociologists argue habits and patterns and behaviour remain consistent over time, it's really hard to change people's behaviour. There's just so much status quo bias kind of built into all of our systems that I do struggle to see where the potential for transformative change comes from, whether in values and beliefs or practices and behaviours.

(M4)

The issue in North America is the consumer norms around vehicle size linked to masculinity, linked to kind of this rural rugged outdoorsy vibe. I mean, the kind of socialised, and sociological side of why you purchase that size vehicle. Of course, that locks you in to massive volumes of petrol purchasing. So the amount that comes out of your wallet is much larger than many of these other countries even though your gas prices are the lowest. So there's this confusing range of signals that are happening.

(F2)

What emerged from the interviews was a sense of, 'wow these are big questions. I am glad you are asking them because I never get the chance to think about it.' The overarching questions of what we are aiming for, what is the path out of this, are not a feature of the discussions people are having as part of their work.

I think my own perspective is that we don't have time to wait for a completely new political or economic system. People have been advocating for reform or overthrow of the system for centuries and so the potential for it happening between now and 2050 or whatever time frame, from the perspective of prudence, it doesn't seem like a very good strategy to rely on that.

(M4)

But that IPCC statement talks about changes to underlying values and things like that and that's so much bigger isn't it? I think it's so much scarier in a couple of ways. It's so much harder to even contemplate. I think because it's so much less clear how we go about doing that and it's a larger ask, but also it's much more of a challenge to the way things are. It's a huge challenge to the status quo, to really talk about you know, changing our norms, and what we value as a society. That entails a massive ask. And so for some people it's just in the 'too hard' basket. So they go about incremental changes, things they can do. Symbolic activities, you know.

(M1)

Whether such a transformation can be achieved. Yeah, I don't know if that could be done smoothly with minimum suffering, or whether it in fact requires disaster where the transformation can only be a phoenix from the ashes thing rather than an organised pre-emptive thing.

(M5)

We could have dealt with climate change through technology and taxes if we had moved fast enough, but not sustainability issues, they require systemic change. I think if we had started to respond in 1990 with the technocratic tools of the right, which is of incremental changes in prices, we probably could have dealt with much of the problem of climate change within the current paradigm. Now what you require is massive intervention of various forms. And so we've completely shifted from this sort of, you know, tweak the system to system change. So the reason we now face, what some people see is this massive interventionist left-wing type agenda, it's because the right chose not to apply the tools it had its fingertips in 1991, 1992, 1993 year after year. I don't think that you could have dealt with it, we could never have dealt with sustainability issues with the tools of the right.

(M1)

The kinds of changes described in the IPCC report are long-term changes, possibly over 2–3 generations at least, and in this time we will see so much climate and ecological loss. But you know, several signs of these transformations already exist. It is not yet clear whether they can coalesce into making macro-shifts happen. But the need for such grand initiatives at transformation has never been greater. As collapses take place, society will be looking for alternatives to the systems that have caused the problems in the first place, and if we don't give these alternatives a chance, we have nothing left to help us get out of crises. And these alternatives are already appearing, already being built.

(M8)

"I wonder if people's response to this is because they can't imagine another world. Like I don't think they're particularly necessarily attached to these systems. I just think maybe it's that lack of ability to visualise what another world looks like.

(F4)

With the contradictions of being green and at the same time, having these world spanning processes of production and transport. I mean, those two things are fundamentally incompatible yet, we don't see that shifting and changing anytime quickly. There's very much a deep attachment to global capitalism, and global supply chains and anything that allows certain companies and industries to be profitable at the service of the consumer. You can't get the mass of society on board, if they literally cannot afford to live, you get spiralling social issues and problems which affect all of us.

(F2)

I feel like there are different views on what dealing with climate change actually looks like and I don't think that we can fully deal with it unless society is restructured.

(F5)

7.2.8 *Waiting for politicians*

The failure of politicians to act is attributed to a variety of mutable factors (rather than any intrinsic failing of liberal democracies themselves). Corruption, conservatism, push back from fossil fuel industries and the agricultural lobby, and a lack of public trust were all identified as factors subverting government action. The answer is for the people to create a sufficiently powerful countervailing force for action.

I don't think we are expecting the government to respond to science. We are expecting the government to respond to people power. The idea that it's an information deficit problem – I think this is a huge issue in the scientific community. It's never expressed anywhere. We sort of have this unspoken assumption that if we generate information our leaders will use that information to make wise decisions in the public good and obviously with climate change there's no better example of that not being the case. So scientists have fundamentally misunderstood the world in which we operate.

Scientists have felt that this was an information deficit problem, that it was simply about having the knowledge. The sciences theory of change is based on false assumptions. And therefore if the knowledge that scientists generate is ignored because it's a game of power and influence, then scientists have to become more powerful, more influential, hence, scientists taking part in vocal civil disobedience to try and, and make ourselves heard, just to try and increase our powers essentially.

(M5)

There's an interesting kind of political discussion there about it's not that the people, the workers, even in the industries, don't want a change. It's that they're not trusting the mechanisms and the actors behind those changes, they feel like they are going to just be left behind. And I think that issue of trust and follow through and wrap around support in a transition for quote unquote real people, that's absolutely essential.

(M2)

There are those of us who have agency, have high missions and agency for change. And there are those who have moderate emissions but very little agency for change. Now I would argue that agency has been deliberately manipulated by a small groups of elites and people have different views on whether it's in journalism or whether it's the politics, I think, they've deliberately managed it.

(M1)

Copenhagen, Helsinki, are cities that can be cited as examples of the direction the world should be headed. The corpse in the room under the table is even in these places, though, you still have these very inequitable global flows of wastes and resources, you know, all that.

(M2)

If you look at the way that we talk about green jobs, for example, we talk about green jobs as being in renewable energy or recycling plants or rewilding or whatever it is. But then there's other incredibly less carbon intensive jobs, like teaching, caring, nursing that no one thinks about. And the reason they don't think about those things is because those professions have always traditionally been undervalued because they're feminized. And so well, what do you do? Do you say 'Well, okay, as part of our green job thing, we're going to take those professions as part of it. And we're all going to value that. Now we're all going to put more of our taxes into it and we all agree'. That's a good thing because those are good professions and because it helps us meet our goals.

(F5)

You know, when politicians for example, look at it they say 'Well people aren't going to vote for, it, it's going to be completely change how they live now.'

(F4)

7.2.9   What's the problem?

Ultimately the changes needed to avert a climate catastrophe depend on what kind of risk we are trying to avert. A different set of responses are needed for a problem that is not very significant, or lies far off into the future, compared to the responses needed for an imminent and catastrophic risk. The chance for 'solving' the problem is in a similar way dependent on the scale and urgency of the risk. Whilst no one suggested it is now too late, nor was there a shared sense that we could escape some level of destruction and upheaval. The key questions are: can we survive, and who is that will survive and relatedly, who gets to define what changes are needed in order to survive?

Those messy processes came up with 1.5 and 2 degrees as appropriate thresholds. And, I mean, they're not safe and I think it's absolutely clear, they're not safe Who should judge whether they're safe or not? Well, those people being impacted, I think they're not bad people to ask whether they think they're safe or not.

(M1)

There are few real solutions to climate change coming from anywhere in the world, including from the global South. But the conversation is being domi-nated by voices, and proposals, from people in rich countries who set forth their own examples. The recent COP was a great example of this – African countries, and their peers, pushed really hard – as they have for decades – to get a mechanism to be paid for the damage caused by climate change. They got the skeleton of this. The narrative in the West was that this came at the expense of action to reduce emissions. Because that is the focus of rich coun-tries and the NGOs that reside there. It did not take on board how people who have a different way of thinking can be valid in their thinking. So when your way of life is being destroyed by weather right now, you might want the focus to be on people paying for their pollution. And you saw this in newspaper headlines and the way things were reported, even in left-of-centre publications like The Guardian The solutions that we are creating have all these assump-tions and worldviews baked in, which tend to be solutions that then continue the status quo and domination of the West. Which won't tackle the climate crisis. The one very solid, big picture, solution that developing countries have pushed is for there to be equal focus on adaptation and mitigation.

(M7)

*I think most fundamentally this liberal norm, where the desires of the individual are the principle around which society should be organised.....
that's not compatible with a finite planet, there needs to be some other organ-ising principle for society and that has to be survival. I use survival as an easy shorthand for just making it, you know, getting through this mess. But I don't know exactly what I mean by it. And what's lost? What's lost is everything we've created over these thousands of years, all that magnificence, all these thousands of different languages, all this knowledge, all of that. It's gone.*

(M5)

It's never too late, you know? And that's the message coming out from the IPCC for the last couple of assessment cycles, at least since the 1.5 report and sixth assessment report, and you know they've been pushing back quite strongly against this narrative that there's a cliff edge and it's 1.5 or nothing or two degrees or nothing.

(M4)

It's like the IPCC as a consensus process is probably conservative. Like we almost have to assume that. Everything, whatever level of urgency or whatever their worst case scenarios are probably conservative estimates because they're consensus processes.

(M2)

We're seeing real climate harm and loss at 1.1 degrees of warming today, and we know that more warming will entail more risks and danger and harm and loss. That's really bad news and it's really tough to package that in a happy way. I guess the hopeful side of it is that if we make the necessary transitions, the fast and fair transition to a fossil-free world, we would have stronger societies, we would have stronger communities. We would have better capacity to respond to climate disasters. Those disasters will be objectively worse. I mean, they will be more frequent. They will be bigger. But if we have a society where we're actually looking out for each other, and using the tools and the creativity that we have, we could be better off. If we used both the technical tools, you know, early warning systems and so on. But also really this social fabric, which is what really helps people in disasters, is your neighbours

(F1)

7.2.10   *It's not just the climate*

All the interviewees recognised that climate change is not just about emissions of greenhouse gases, but an array of interlocking issues, including democracy, justice, racism and power.

What would it actually look like to move towards a social ecological future that puts meaningful climate action at its core in a way that also is trying to address inequality? The colonisation concerns, gender equity, good governance, to demonstrate that we're moving towards environmental sustainability in ways that also ensure greater equity, that address both global North-South inequalities but inequalities within societies, that addresses gender inequality and racism.

(M2)

Who gets to sit at the table and determine value. This is where we get into transformative issues of representation. Who sits at the table, who makes decisions at those tables. Are they consultative symbolic panels, or do they have power to actually change processes? So there's a lot that needs to be done on reshaping democracy to be more representative, more participatory, more direct. If you want change at a global and national scale, that means digging into what it would mean for those groups to feel heard, and what's really at the core of that. And sometimes it's racism and sometimes it's a deep sense of like my position in the world has fallen relative to others. I think unpacking that is where we get into the language and ideology of what are you free to do and at what cost and who has that freedom?

(F2)

Is it possible to have a system built on extractive ever-growing capitalism that shifts money to the people at the top permanently, and at the same time tackle climate change? Yeah, that's where it gets sticky. I think it's hard to see how those value systems would operate in parallel. Pushing all these things, whether that's politics, power, social change has to be part of tackling the climate crisis, even if that doesn't mean that we build a utopian world. Let's say, for the sake of argument, it is possible to get to very low emissions by 2030. Let's say we pull that off. That doesn't mean that we have utopian political or social systems and in some versions of that it will be worse.

(M6)

Recently I heard somebody talk about disaster justice and others were frustrated like, why do we need all these different justices? Somehow it's all about social justice ultimately. But it's just that we need these types of labels because people are not willing to look outside their kind of interests. But I think it's all about this, the inequality, the unfairness. And you know, it's rooted in, obviously racism and capitalism and capitalism and racism in league together and colonialism and these power issues. I can't see how there is any way we could possibly transform or rather, I should say how we could possibly cope with the changing climate without transformation and transformation to me is fundamentally about this.

(F6)

7.3   Conclusion

This analysis reveals how difficult it is to imagine an alternative society. Even for these interviewees – who are under no illusion about the threat we face, who spend their working day thinking and communicating about how to avert disaster, and agree with the IPCC statement about the need for fundamental change – it is almost impossible to articulate the contours of a radically different society. Instead, we find ourselves retreating into quietism (my work doesn't allow me to express my real concerns and wishes), smallism (this stuff is too big, we need to focus on more manageable and tangible wins) and voluntarism (we must create the conditions which allows individuals to choose to change their behaviour).

Despite the difficulties of imagining alternatives to a society built on the principles of free market liberalism, interviewees were able to find reasons to be hopeful. The ordering principles of the Sustainable Development Goals and the concept of sustainability itself both challenge existing ways of being in the world that a narrow focus on emissions reductions doesn't. The barriers to the realisation of sustainability as a new ideological north star include the reactionary nature of vested interests, structural factors making it difficult for people to enact sustainable behaviours, countervailing social norms, and the lack of strong political leadership. Added to this is the urgency of the climate crisis, meaning there is insufficient time to try and build a different future (though for one interviewee it was the very lack of time that meant it was now too late to continue waiting for technological interventions. Radical social and political change is the only remaining option).

Most interviewees were also hopeful for the future because they held a belief in the ability and desire of people as individuals to change. Whilst visible signs of this appetite for change were few and far between, that didn't mean that at a subterranean level, attitudes aren't changing. The state itself was rarely directly identified as the agent of change. Any profound social change that does occur is generally thought to arise outside of the state, the state will not be the driver of radical change.

Ultimately, the responses to the questions asked revealed a shared agreement with the IPCC statement regarding the need for fundamental change to how we relate to the world and each other in order to tackle climate change. It was widely felt that those working in the arena of climate change should not seek to define what that new world looks like. To do justice to the diversity of human experience means recognising people will be working towards a multitude of different futures. People were identified as key to bringing about this change, acting either individually or collectively. Interviewees were not able to predict the social tipping point that will usher in this change. All that campaigners and communicators can do is to continue doing the work of creating the conditions that will allow sufficient numbers of people to voluntarily choose to change their behaviour to create that social tipping point.

8   Conclusion

What future?

8.1   Is there a there there?

The four years it has taken to write this book has been witness to a growing chorus from powerful global institutions, calling for fundamental change to our way of life in order to avert climate catastrophe. These declarations are coming from discursive communities held together by liberal norms. Are these institutions saying that liberal norms must now be abandoned, and replaced with a whole new set of values and ideologies? If so, this would likely be the first record of a ruling ideology seeking to dissolve itself, with no revolution or conflict driving that need.

As this book has shown, when one digs a little deeper, it becomes clear the transformational change described is not very transformational at all. The interviews discussed in Chapter 7 demonstrated that there is no alternative society waiting and ready to replace the world destroying liberal system. As Foucault and others have explained, for the first time since the reformation, there are no longer any significant oppositions within the thought world of the West.

There is no shared awareness of and desire for the overthrow of liberalism. Liberalism has no challengers. Geuss (2022) suggests we are happy to be complicit prisoners in the intellectual straitjacket of liberalism because liberalism offers the fantasy of agency in a world in which we have lost any real agency. It is this idea of agency that we seem unable to surrender, as it is this agency, expressed by voluntary adoption of low-carbon behaviours, which is widely held to be a necessary, if not sufficient, element of humanity's salvation.

The interviewees made a spontaneous connection between the individual and consumerism. It is in our choices of what to buy that we create the private world we want, rather than acting together politically to build a different shared world. It's just easier that way. Democratic politics was once upon a time designed as a check on the undemocratic economic sphere, an effort to prioritise the public over the private. But honestly, who can be bothered any more? As a consumer, the focus of activity is on what *I* want. As a citizen, our actions are directed to what *we* want (Sagoff, 1998). It is only as a consumer that the individual is sovereign, and hence since this is the only arena in which we feel we have any control, it is here we direct our activities, our search for meaning, rather than the collective or the social. But even this sense of control is an illusion in a world where

DOI: 10.4324/9780429463488-9

the annual global advertising spend is somewhere north of $700 billion (Dentsu, 2022). The relatively miserly funds and resources available to those seeking to challenge this hegemony make for a very one-sided contest. People are not voluntarily going to abandon consumerism, their joy of buying and travelling, in order to avert ecological collapse. Urry has written extensively about mobility and freedom in modernity. Whilst in 1800 in the US, people travelled 50 m a day they now travel 50 km a day (Buchanan, 2002, cited Urry, 2010: 89). This has happened as a result of social practices which presuppose huge increases in the distance and speed at which people and goods travel, meaning life is now "mobilized" (ibid.). Our societies are built to facilitate this mobility and either societies are reconfigured, or we find new ways to fuel this mobility, or we just carry on as we are. No matter how bad the impacts become, even if people wanted to give up the pleasures of consumerism, they simply won't be allowed to.

The "individual" is the foundational unit of human existence in the liberal universe. Campaigning and research are easier when you target the individual rather than the system. Unable or unwilling to challenge the capitalist state and corporate power, campaigners and communicators instead turn to the individual – weak, atomised, manipulable – as the agent of reform and lifestyle adjustment. This focus on the individual and their beliefs and attitudes as the locus of change is so pervasive that we are left unable to engage with the kind of systemic change described by the IPCC. It simply isn't a part of the day-to-day job for researchers, campaigners and communicators to imagine alternative societies.

Marxists such as Christopher Caudwell have argued that the human world is social, all the way down. In our actions, we are simply carriers of our social position. That is all there is. The individual is an abstraction from the only reality, our social existence. The primary role of the individual in capitalism is not not to drive social change, but adjustment, adjustment to the demands of the system (Margolies, 2018). In this process of adjustment, the individual, through their discourses and practice, reproduces the system. Any agency is limited to the realm of the prosaic.

Our ability to act is dependent on the social networks of which we are a part. These social networks and practices are themselves historical phenomena, the artefacts of many years of sustained social, economic and political practices. Today in the West, our social relations are defined by liberal ideologies, and we seem incapable of imagining anything different because all our social relations, institutions, cultural discourses and ideas of what is good are defined by that ideology. It is liberalism as far as the eye can see, the heart can feel, the mind can imagine. There are very real limits to the extent to which one can stand outside the prevailing norms of one's social and historical situation. Transplant a hunter gatherer to the home counties of the UK and the likelihood is that their hunter gatherer skills would be of limited use. They would not be able to take advantage of the material abundance surrounding them. Their inability to survive would be more to do with cultural and social factors than the natural environment.

In a liberal economy our goal is to compete with other individuals for economic resources. It is no longer a class-based competition, or at least, it is not recognised as such by the majority of subaltern actors. Our social environment is one of individuals pitted against other individuals in the market place of life chances. The material environment has been built to facilitate that goal. Now however, climate change means we must learn how to sustain that social and material environment whilst limiting emissions of greenhouse gases. The essay 'The Tragedy of the Worker' (Salvage Collective, 2021) explains that fossil fuel energy is itself inert without human energy; the capacity of fossil fuels to create profit is dependent on human energy. Capitalists unite the two, fossil fuel energy and human energy, seeking to extract maximum energy from both in order to maximise profits. This means that the goal of asking people to act in low-carbon ways within the framework of capitalism amounts to getting workers to do the work of creating the conditions for the replacement of fossil energy with other forms of energy, so that capital can continue extracting their labour in pursuit of profit. No longer can the tired worker rest in her own time. She must now use her spare time to create the conditions for her own continued low-carbon exploitation.

8.2   The limits of the individual in a world of limits

Let us consider two ways of apprehending the world. The first asks us to trust in something called the individual, who will, in response to external signals (price, symbolism, experience of climate extremes) voluntarily choose to change their behaviour. In this world there is nothing beyond individual will, no meaning, no structure, nothing bigger than the individual. The pluralist ideation we find in liberal societies supposes that if sufficient numbers of individuals make low-carbon choices, then the activities of businesses and the state will change to accommodate these new patterns of behaviour. The second option is to turn to the state or some other entity or structure bigger than ourselves to tell us what to do, to make the decisions for us. This structure will prioritise the collective, the social, over the individual. Our actions will be guided by a long-term perspective, rather than an eternal "now". This ability to adopt a longer-term view, as Burke identified, is what makes the state the only form of organisation that can supplant the existing destructive patterns of activity. A state not captured by bourgeois interests can be a foil to the trend for each new generation of

> life-renters…unmindful of what they have received from their ancestors, or of what is due to their prosperity, should act as if they were the entire masters who…think it amongst their rights to cut off the entail, or commit waste on the inheritance, by destroying at their leisure the whole fabric of their society.
>
> (Burke, cited Turner, 2003: 81)

Liberalism, or what Curtis White calls the 'Middle Mind' (2003) imagines a place in the middle of these two world views, between hyper-individualism and authoritarian state control. This centrist narrative claims that it is not either/or. The social is generated through an interplay between the system and the individual. An effective response to climate change requires both the system and the individual to change concurrently, in a mutually reinforcing and non-conflictual process. Power relations are largely absent from this discourse, except to highlight that individuals acting as individuals have agency and power. That these changes seem no closer to happening now than thirty years ago does not seem to dishearten the holders of this opinion.

To agree with the IPCC statement that we need a change in our values requires we reject this idea of a middle ground. Watt, in his discussion of the rise of the novel, shows how the emergence of economic individualism in the 17th century as the overriding value in capitalist societies has, and must always, displace other values (Watt, 2000). Or, as Cohen put it, when Homo Economicus enters the room the moral human leaves (Cohen, 2014).

The dominant values of the economy in the West are characterised by an attitude of 'what can I get away with?'. This 'what can I get away with' economic value extends to capitalism's attitude to the natural environment. The climate targets designed to limit warming to 1.5°C are the economic ethos translated directly to the natural world. How much warming can we, the wealthy societies in the West, get away with? This calculation is a tragedy, sold as a solution.

The "mushroom" view of the world, which suggests that the required change is building, out of sight underground, ready to break through at any minute, could yet prove true. But at this moment, it is difficult to imagine people will individually and voluntarily jettison all the norms, shared understandings and common practice of their surroundings and choose to live a completely different life at odds with or unconnected with those norms. Even when given the chance in elections to usher in a new, just, and sustainable world with no more than a tick of a box, people choose to stick with the status quo. The Swiss have recently rejected a constitution designed to help the country meet its climate targets (BBC, 2021). Chileans also recently voted against a new constitution designed to implement radical new social and environmental goals (BBC, 2022). It appears Hegel was on to something when he described the people as that part of the state that does not know what it wants (Hegel, 2008).

8.3   You shall have no other gods but science

Science has crowded out the possibility of approaching climate change as an issue to be addressed through a revision of values. Science is our value, instrumental rationality is the assumed ideal of Western civilisation. When the cultural critic de Zengotita asks 'Isn't it, in the end, the betrayal of the liberal ideals of the Western enlightenment that make you angry, not the ideals themselves?' (de Zengotita, 2003) the betrayal he speaks of is the primacy given to instrumental rationality, the logic of efficiency, profit and performance. This is what White calls 'bad enlightenment', rather than the 'good enlightenment' of emancipation, human self-realisation, creativity and democracy (White, 2003: 121). In this 'bad enlightenment', the dominance of instrumental rationality has created a technocracy in which those who govern justify themselves by appeal to technical experts who, in turn, justify themselves by appeal to scientific forms of knowledge (Roszak, 1969).

The future has become inseparable from projections of scientific and technological advance (Crary, 2022: 55).

Our technocracy is the outcome of attributing the whole responsibility of producing facts to a happy few (Latour, 1987: 174). The global science associated with climate change research, has been described as regulatory science – a mutual construction between scientific research and policy needs (Szerszynski, 2010: 19). Policy needs dominate that relationship, as made clear by the creation of the IPCC to provide 'policy relevant science'. It is a process open only to those with the correct credentials, a happy few.

Now, this may be all fine, and I will join in the obligatory note of congratulation to science for the medical advances that I enjoy today and which have extended my old age by a few years. But the point is, science can't tell us what to do about climate change. We have made ourselves dependent on an impotent and, in the case of climate change, absent god. As a recent essay on the work of the Jewish philosopher Günther Anders records, the more technology allows us to do the more it necessarily also does "without us" – that is, without our active comprehension, involvement and thus, ultimately, also without requiring our explicit consent or generating concern. Increasingly, our things don't seem to need us anymore (Borowski, 2022). This undermines our agency and limits the sphere of democratic control.

Liberal voices have always presented environmental issues as problems needing a scientific response, not a change in values. An Independent newspaper editorial from 1989 encapsulates this opinion: 'Brown rice and sandals do not amount to a cure. If we are to treat environmental problems successfully, it stands to reason that our approach must be informed by scientific understanding.' (Independent editorial 25th September 1989, cited Yearley, 1992). There is a direct and unbroken line from this editorial to more recent remarks from the Swedish Deputy Prime Minister quoted previously, that 'we must trust to innovation, the speed of which continues to amaze' (The Guardian, 2018). Or, to put it another way, we must trust in the lord's wisdom, for he moves in mysterious ways. Those of us not of this faith must face the question posed by Winner nearly 40 years ago. 'What shall we do when faced with the inadequacy of our measurements?' (Winner, 1986: 127). It has become impossible to ignore the suspicion that the single-minded pursuit of freedom and efficiency must inevitably lead to disaster (Lila 2015, cited Betty, 2017: 124). We are lost, we do not know where to turn, what direction to head in, what it is we want. We need help, and we must look beyond the individual for that help.

8.4 We can't do this on our own

Willa Cather's novel, *The Professor's House*, first published in 1925, features a long monologue in which the main character laments that whilst science has given us 'a lot of ingenious toys' these same toys 'take our attention away from the real problems, of course, and since the problems are insoluble, I suppose we ought to be grateful for the distraction.' Life, the professor claims, had meaning for everyone when life was a religious drama in which everyone has a role in the battle between good and evil. King and beggar alike were united in this drama. In being surrounded by and part of the 'pomp and circumstance' of religious rituals humans found the only happiness they have ever had. Participation in religious ceremonies gave people a feeling of dignity and purpose, and every act had an imaginative end (Cather, 2006).

The real problem that science cannot address is the search for meaning in an increasingly godless, and nihilistic world. Jacques Ellul and Erich Fromm, amongst others, trace the development of individualised social relations to the breakdown of mediaeval society. The collective nature of mediaeval society was, in these accounts, superseded by the atomisation of human relationships, which Ellul identifies as a necessary precursor to the development of industrial society. This atomisation was inevitable following the loss of the support and sense of belonging provided by mediaeval social and religious structures, and was replaced with a developing sense of the self as an individual. Fromm and Ellul both see this individualism as offering a freedom from the ties of community and tradition. But, they argue, this freedom left people feeling isolated, afraid and in need of support and a sense of belonging. There was nowhere left to turn for this support but the newly emerging nation state. Consequently, the nation state becomes a type of secular God, the giver of laws, close at hand yet remote with powers of life and death. But the state-god is unable to help us answer the key questions asked by the citizens of this new world, namely 'Is my life amounting to something?'(Giddens, 1991). Absent God sitting in judgement whether or not one's life has been well lived, and how do we give our lives meaning? For Giddens (1991), the construction of meaning and identity is an ongoing task, a matter of asking ourselves each morning, who shall I be today?

And, as history has shown, the technocratic search for perfect efficiency cannot sit alongside alternative values. It is all in, or nothing. Any criticism of this dependence on science and technology is easily brushed aside as a hopeless and reactionary desire to return to the caves, whilst our goal should instead be to keep the lights on. And yet, the challenge remains – we need stories we can all feel a part of and provide meaning beyond merely limiting hope to a future exactly the same as today, with fewer emissions of greenhouse gases. And science, as a highly specialised and technocratic discourse accessible and attractive to only a small caste, cannot provide those stories. With organised religion in the West in a continuing decline, and the remaining religious activity becoming ever more fractured, the only option we have with such little time left is to turn to the state.

Spinoza, writing in 1670, hoped for a society in which rational, virtuous and free individuals acted purposively and morally as a result of their individual grasp of the truth of the world, rather than pursuing collective goals under the yoke of religious doctrine. In such a world, there would be no need for the state (Nadler, 2011: 19). We seem as far from such a utopian model of the ideal human today as in Spinoza's time, and so we must remain dependent on the state for creating a new set of values. The issue is less one of humans as essentially irrational, at least as the term is normally used. It is more that we are being guided to act in terms defined by instrumental rationality, rather than a value rationality. Value rationality is to be guided by a moral or ethical end. It is a singular and permanent objective.

Value rationality is absolute, an overriding principle to guide action. Instrumental rationality on the other hand, aims only to satisfy a particular desire through the most efficient means possible. All other considerations are secondary to this. It is this value rationality which needs coordinating and promoting by an entity beyond ourselves, an entity such as the state.

Liberal climate change policy pays little heed to any value rationality beyond the reproduction of the liberal hegemony. The non-negotiable demands of liberalism, in meeting the non-negotiable physics of the climate system, has generated a discourse bounded by the five climate guardrails explored in this book. These guardrails prioritise the reproduction of liberalism over an effective climate response, and preclude the possibility of imagining a world built on different values. Instead, we are presented with visions of the future that have all the features of a Xanadu like hallucination, a pleasure dome where as of yet non-existent technologies remove the need to look beyond liberalism for answers to climate change, whilst climate impacts have disappeared. The various deceits underlying the policy promises which define the guardrails of liberal climate change have been met with hardly a whisper of protest because no one really understands the sacrifices being demanded in order to save liberalism. It is clear that liberalism's pyrrhic victory is assured because of its success in co-opting climate change.

We are unlikely to build anything to replace what currently exists without the organisational structure and the force of conviction of a religion (Morrey, cited Betty, 2017: 48). The etymology of the word religion is rebinding, *religare,* binding people together under a shared faith (ibid.: 60). Unfortunately, Western organised religion has failed us. The Catholic church has promoted dogma over social justice, whilst Protestantism has demoted the social function of christianity by personalising the relationship between people and god (White, 2003: 86), ironically leading to the rise of individualism.

Without the building and maintenance of an alternative value system by an overarching idea and power structure, the only way things can ever change is through individual change. Currently, neither religion nor the individual is able to guide us out of the climate impasse. The only existing actor able to act urgently is the state; the state is the most powerful source of collective action in history (Lieven, 2020: xv). Burke (1791) spoke for those who felt concerned by what they saw as the excesses of the Enlightenment (e.g. the French Revolution) when he said 'Society cannot exist unless a controlling power upon will and appetite be placed somewhere, and the less of it there is within, the more of there must be without.' The only problem with surrendering our few remaining freedoms as consumers to the state is that, just like organised religion, the liberal state has failed us. This failure can be understood with reference to what Trotsky called 'the real history of liberal democracy – successive attempts of the capitalist class to emancipate itself from the working class' (Trotsky, 1919). Democracy 'is a facade behind which bourgeois hide their power and we will need a dictatorship until the bourgeois have disappeared as a class' (ibid.). This is not a universally popular claim with regard to discussions of climate change and democracy. Fischer (2017) alongside Mann and Wainwright (2017) fears climate change will lead to exactly this kind of authoritarianism, but what they fear is actually a fusion of authoritarianism with liberal values, wherein as impacts worsen, elites retreat to their castles in the sky whilst the state is employed to protect the elites from the ever more desperate masses. Hence, climate change is an issue that can only be solved with more democracy, not less (Machin, 2013; Martell forthcoming). There is, however, little democracy to be had in a highly technological and technocratic society, where the demands of technology must come first. What little democracy and freedom exists is simply the leftovers after the important decisions directed at the maintenance of a technological society, the needs of finance and trade, and foreign policy strategy have been made.

The belief that the people are a mindless mob who can be easily manipulated is a long held key principle of bourgeois liberalism (Thorpe and Gregory, 2010: 296). And the people have been successfully manipulated for a long time. The maturity needed to be able to engage in decision-making comes from being given the chance to do so. The people have not had that chance, and instead spend much of their conscious time in "bullshit jobs" (Graeber, 2018) which further erode our capacity to act in an alienated world. Our daily decisions are not how to maintain an ecologically sensitive and socially just world, but whether or not there will be time to get to the supermarket on the way home tonight. People do not want to give up their spare time to have greater involvement in decision-making, at least not as regards climate change. Invitations to participate in climate assemblies in the UK, even with the promise of payment, elicit a response rate of around 5%.

Lenin identified in the bourgeoisie a lack of faith in the people, a fear of their initiative, and trepidation before their revolutionary energy instead of all round unqualified support for it. The fear and shock felt by European policy elites following the Gilet Jeunes protests in France has led to a new interest in discussions of a "just transition", and new funding opportunities for the organic intellectuals to work out what actions are needed to prevent such spontaneous working class opposition ever happening again.[1] This fear is also apparent in the design of climate assemblies, wherein the issues opened up for debate are carefully curated in advance by the participant's betters.

Lenin saw that humanity's future depended on acceptance of a truth antithetical to bourgeois liberalism – there is no middle course, it is socialism or barbarism. Liberalism is a commitment to barbarism and any effort to build a socialist future in the West will meet with the same levels of barbaric and sustained violence which faced the Soviet Union from the very start of the revolution (Fogleson, 1995). The 1917 Russian revolution was a conscious intervention to end imperialism and class exploitation. It was a seismic act, a truly historical moment, the reverberations of which echo still today. We desperately need a similar historically significant social revolution today, one which recognises the primacy of the state as the first and prime agent of the people's interests during this shift to a world of new values.

The gradualism and long-termism of social democracy has run its course. It has no energy, no vitality, no life left in it. It is a system that has sacrificed passion for efficiency, a quiet, orderly shuffling to the world's death. Passion alone is the cause of all historical events (Hegel, 1956, cited Whyte 1973: 107). The state, as an ideal, mediates between reason and passion (ibid: 108) and removed from serving bourgeois interests, offers the hope of using that passion and energy that is found in the subaltern classes to create historical change. The death of liberalism is the means of a transformation to a higher level of consciousness than the life that led up to it (Whyte, 1973: 118).

The state can also mediate between the local, national and international. Ostrom offered a detailed analysis of this balance between the central and the local. Ostrom's principal interest was in how institutions worked or failed to sustain collective resource use. Ostrom noted that self-governing entities exist at a variety of scales and can be found in both the public and private sphere. The key question for Ostrom was: 'How can fallible human beings achieve and sustain self-governing ways of life and self-governing entities as well as sustaining ecological systems at multiple scales?' (Wall, 2014).

Ostrom's answers to these questions were influenced by a range of classical liberal economists such as Joseph Schumpeter and Friedrich Hayek, with whom she shared a scepticism about governing through centralised state power. Her scepticism stemmed from the belief that state actors do not have perfect knowledge of the issues being faced by communities geographically and culturally distant from policy elites. But, Ostrom also recognised that local knowledge on its own is insufficient because it is not always correct. Ostrom's experiments and research showed that in real life there are often high levels of co-operation, communication and information sharing. This co-operative set of social relations allows people to work collectively to achieve long-term optimal outcomes for the resource users. Ostrom was not writing as an anarchist, but was highlighting how centralised knowledge and local knowledge can together offer a means of sustainably managing our common resources (Wall, 2014).

8.5   A peasant prospect

Our consciousness is historically determined. It is the weight of history, expressed through every facet of the present, which makes it so difficult to look beyond existing norms for answers to climate change. To argue we need to stop trying to fix the world, stop trying to make everyone a member of the bourgeois, that in fact, we need a world that becomes more working class and less middle class (whatever that means in different parts of the world) seems shockingly naive and unrealistic because it seeks to push against the weight of centuries of efforts to remake the world according to the demands of market forces. But the greater part of human knowledge is not scientific or technological at all, merely conventional, practical and pragmatic knowledge (Croce, cited Whyte, 1973: 382). This is not the kind of knowledge the bourgeois went to university to gain, that allows the bourgeois to distance themselves from the working class.

Faced with the question 'Should we spend our time working to improve techniques of risk analysis and risk assessment? Or should we spend time more directly to find better less risky ways of living?' (Winner, 1986: 175) this author opts for the latter, and extends that proposition to suggest, with mild irony, that only apathy, uninventiveness and inertia can save us (Forster, cited Passmore, 1978: 40). This low ambition ideology is the kind of life liberalism has sought to destroy wherever it encounters it.

What is there to fear in such a prospect? Not a reduction in human happiness it would seem. The Harvard University's Study of Human Development Research programme has been running for 84 years and has found, repeatedly, that what leads to happy and healthy lives is close relationships, much more than money or fame (Harvard Gazette, 2017). Happiness depends ultimately on comparisons that each person establishes with a reference group, and what we seek is some measure of equality with our reference group (Cohen, 2014: 7). A more equal world, a less grasping world, will be a happier world. The fundamental basis of human nature is sociability and it has required a whole world of inventions against nature to prevent men from living together (Michelet, cited Whyte, 1973: 151). A world where the collective matters more than the individual is the world we should promote if we want to live within climatic limits. People who have adopted lower carbon life-styles cited reasons of social justice, community, frugality and personal integrity as primary motivations, rather than the environment. Climate change was not seen as particularly interesting by the research participants (Howell, 2013). The means of building a low-carbon world then is not to focus on emissions, but on repairing the shared sense of meaning and identity taken away from us by liberalism.

This chapter hasn't spent much time detailing possible alternative futures because it is a pointless exercise, no such futures will emerge in time. In 1967, Martin Luther King Jr realised tinkering around the edges of society was a waste of time, that what we needed was 'a revolution in values.' Today social cohesion is much weaker than it was fifty five years ago. There is no shared vision, no 'us'. And yet one still hopes, one continues fighting. What else is there to do? In 1888, the historian Michelet wrote 'My very heart bled in contemplating the long resignation, the meekness, the patience, and the efforts of humanity to love that world of hate and malediction under which it was crushed' (Whyte, 1973: 154). Yet, even after 150 years of the iron boot smashing into the human face, we still dream of something better. It is only in the bravery of those unwilling to join the mute shuffle to extinction that any hope lies. Tocqueville noted it was those areas that had some reform that the clamour for radical revolutionary change the loudest. When offered no hope of release people accept their fate, but just a taste of a better alternative, and the dam breaks (Whyte, 1973: 217). We ought to continue searching for a chink of light from the other world in the darkness of instrumental rationality, and let love, compassion and humility shine through, only then may we be able to limit the ecological genocide underway. That means finding a way to understand and know the world beyond the guardrails that serve only one God, liberalism.

Note

1 See, for example, the recent Fair Energy Transition For All project. "This is the challenge: to further the European Union's energy transition without a public backlash that would not only jeopardise climate goals but risk the cohesion of our societies. It is a challenge taken up by a consortium of philanthropic foundations, moved to act, notably, after chaotic street protests against fuel duty hikes shook France in 2018–2019." Final

Report. Fair Energy Transition For All. How to Get There. (2022) Retrieved from https://fair-energy-transition.eu/wp-content/uploads/2022/11/FETA_SYNTHESIS_final.pdf.

References

BBC. (2021). *Swiss Voters Reject Key Climate Change Measures*. Retrieved from https://www.bbc.co.uk/news/world-europe-57457384

BBC. (2022). *Chile Constitution: Voters Overwhelmingly Reject Radical Change*. Retrieved from https://www.bbc.co.uk/news/world-latin-america-62792025

Betty, L (2017). *Without God: Michel Houellebecq and Materialist Horror*. Pennsylvania: Pennsylvania State University Press.

Borowski, A. (2022). 'Philosopher of the apocalypse'
. *Aeon*. Retrieved from https://aeon.co/essays/gunther-anders-a-forgotten-prophet-for-the-21st-century

Burke, M. (1791). *Letter to a Member of the National Assembly of France*. Retrieved from https://quod.lib.umich.edu/e/ecco/004804929.0001.000/1:3?rgn=div1;view=fulltext

Cather, W. (2006). *The Professor's House*. London: Virago.

Cohen, D. (2014) *Homo Economicus: The (Lost) Prophet of Modern Times*. Oxford: Polity Press.

Crary, J. (2022). *Scorched Earth. Beyond the Digital Age to a Post-Capitalist World*. London: Verso.

Dentsu. (2022). 'Dentsu ad spend report predicts continued growth through 2022 despite global economic turbulence'. Retrieved from https://www.dentsu.com/news-releases/dentsu-ad-spend-forecast-july-2022-release#:~:text=The%20latest%20dentsu%20Global%20Ad%20Spend%20Forecast%20points%20to%20a, adjusted%20growth%20forecast%20to%208.7%25

de Zengotita, T. (2003). 'Common ground: Finding our way back to the enlightenment'. *Harper's Magazine*. Retrieved from https://www.thefreelibrary.com/Common+ground%3a+finding+our+way+back+to+the+enlightenment.+.-a099601601

Fischer, F. (2017). *Climate Crisis and the Democratic Prospect: Participatory Governance in Sustainable Communities*. Oxford: Oxford University Press.

Fogleson, D. (1995). *America's Secret War against Bolshevism: U.S. Intervention in the Russian Civil War, 1917–1920*. Chapel Hill: University of North Carolina Press.

Geuss, R. (2022). *Not Thinking Like a Liberal*. London: Belknap Press.

Giddens, A. (1991). *Modernity and Self-Identity*. Cambridge: Polity Press.

Graeber, D. (2018). *Bullshit Jobs: A Theory*. London: Allen Lane.

Harvard Gazette. (2017). *Good Genes Are Nice, But Joy Is Better*. Retrieved from https://news.harvard.edu/gazette/story/2017/04/over-nearly-80-years-harvard-study-has-been-showing-how-to-live-a-healthy-and-happy-life/

Hegel, G.W.F (2008). *Outlines of the Philosophy of Right*. Oxford World Classics. London: Oxford University Press.

Howell, R. (2013). 'It's not (just) "the environment, stupid!" Values, motivations, and routes to engagement of people adopting lower-carbon lifestyles'. *Global Environmental Change*, Vol. 23, No. 1, 281–290,

Latour, B. (1987). *Science in Action*. Cambridge: Harvard University Press.

Lieven, A. (2020). *Climate Change and the Nation State*. London: Allen Lane.

Machin, A. (2013). *Negotiating Climate Change: Radical Democracy and the Illusion of Consensus*. London: Zed Books.

Mann, G. and Wainwright, J. (2017). *Climate Leviathan: A Political Theory of Our Planetary Future*. London: Verso.

Margolies, D. (2018). *Culture as Politics Selected Writings of Christopher Caudwell*. New York: Monthly Review Press.

Martell, L. (forthcoming). *Alternative Societies. For a Pluralist Socialism*. Bristol: Bristol University Press.

Nadler, S. (2011). *A Book Forged in Hell*. Princeton, NJ: Princeton University Press.

Passmore, J. (1978). *Science and Its Critics*. London: Duckworth.

Roszak, T. (1969). *The Making of a Counter Culture*. New York: Doubleday.

Sagoff, M. (1998). *The Economy of the Earth*. Cambridge, MA: Cambridge University Press.

Salvage Collective. (2021). *The Tragedy of the Worker. Towards the Proletarocene*. London: Verso.

Szerszynski, B. (2010). 'Reading and writing the weather'. *Theory, Culture & Society*, Vol. 27, Nos. 2–3, 9–30. doi: 10.1177/0263276409361915.

The Guardian. (2018). *To Lead on Climate, Countries must Commit to Zero Emissions*. Retrieved from https://www.theguardian.com/environment/2018/apr/17/to-lead-on-climate-countries-must-commit-to-zero-emissions

Thorpe, C. and Gregory, J. (2010). 'Producing the post-Fordist public: The political economy of public engagement with science'. *Science as Culture*, Vol. 19, No. 3, 273–301.

Trotsky, L. (1919). *The Principles of Democracy and Proletarian Dictatorship*. Retrieved from https://www.marxists.org/archive/trotsky/1918/xx/principles.htm

Turner, M. (2003). *Reflections on the Revolution in France*. New Haven, CT: Yale University Press.

Urry, J. (2010). 'Sociology and climate change', in Carter, B. and Charles, N. (eds). *Nature, Society and Environmental Crisis*. Oxford: Wiley-Blackwell, pp. 84–100.

Wall, D. (2014). *The Sustainable Economics of Elinor Ostrom. Commons, Contestation and Craft*. Oxford: Routledge.

Watt, I. (2000). *The Rise of the Novel*. London: Pimlico.

White, C. (2003). *The Middle Mind. Why Consumer Culture Is Turning Us into the Living Dead*. London: Penguin.

Whyte, H. (1973). *Metahistory: The Historical Imagination in Nineteenth Century Europe*. Baltimore, MD: John Hopkins University Press.

Winner, L. (1986). *The Whale and the Reactor. A Search for Limits in An Age of High Technology*. Chicago: University of Chicago Press.

Yearley, S. (1992). *The Green Case: Sociology of Environmental Issues, Arguments and Politics*. Thousand Oaks, CA: Taylor and Francis Ltd.

Index

Note: Page numbers followed by "n" denote endnotes.

For Product Safety Concerns and Information please contact our EU
representative GPSR@taylorandfrancis.com
Taylor & Francis Verlag GmbH, Kaufingerstraße 24, 80331 München, Germany